HUMAN ANATOMY

HUMAN ANATOMY

Stereoscopic Images of Medical Specimens

FROM THE COLLECTION OF THE VROLIK MUSEUM

JIM NAUGHTEN

Hoop Editions

PRESTEL
Munich • London • New York

CONTENTS

Anatomical Collections: The Dead Teach the Living

Laurens de Rooy

Who is responsible for the billions of liquid specimens that are displayed in the natural history and anatomy museums of the world? In other words, who discovered the preservative qualities of alcohol? Some say it was Anglo-Irish scientist and philosopher Robert Boyle (1627–91) who first put a dead animal in a jar with alcohol, but this is disputed. We do know that in 1662 British physician William Croone (1633–84) appeared before the Royal Society in London, showing two specimens of puppies preserved in alcohol.[1] It is the earliest written record using this preservation method. Around the same time, Dutch anatomist Frederik Ruysch (1638–1731) started to create his first anatomical preparations, but whether these were preserved in liquid or not is unclear. He did, however, set the stage in 1685 for all later anatomy museums by opening the doors of his private collection to the public. Ruysch had anatomical preparations in alcohol on show, and the museum became an attraction. In 1697, Peter I, Emperor of Russia (1672–1725), visited and was so impressed by what he saw that he bought the entire collection in 1710.

Ruysch's preparations were and still are the most famous of their kind. He was the first anatomist to inject blood vessels with a red wax-like substance that could travel up to the capillaries in, for instance, the skin of the face, thus creating the appearance of blushing as if still alive. This effect was heightened by the use of lace bonnets and collars to cover up the area where the neck or head had been cut off.

Ruysch used this technique to preserve the results of his own research, and his collection served as proof for his theories as well as teaching material. A collection of this kind during the seventeenth century also displayed a certain status: that of a cultured man of medicine and surgery, a craftsman or even an artist of anatomy. Ruysch did not just use his craftsmanship to highlight his knowledge, but also to evoke emotion in the spectators, making them wonder at all kinds of things: the miracle of creation, the inevitability of death, the aesthetics of the body. He attempted to please the viewer.

These anatomical preparations started as parts of living things, but became material culture; objects. As objects they evolved into means to demonstrate, or better, to communicate knowledge, theories, the essence of a disease, aesthetics and emotions. And, from that point on, this has been the purpose of all anatomical museums or collections.

By the end of the seventeenth century and throughout the eighteenth century, many more anatomists and surgeons started their own collections, and more and more of these contained large numbers of liquid specimens. These collections were, of course, anatomical, but at the same time they conformed to the enlightenment culture of classification at that time. To quote Sam Alberti, former director of the Hunterian Museum in London: 'Anatomy was a broad church, including human and animals (and plants), diseased and healthy – but at its core was the dissected, preserved body.'[2] In the UK, anatomy museums thrived in the second half of the eighteenth century. 'This abundance can partially be explained by shifts in the way anatomy was studied in the eighteenth century: with more dissection, with more objects. Surgical anatomy relied on in-depth knowledge of the dead body, which could only be partly provided by fresh dissections.'[3] Thus, sizeable collections were found in several places. Anatomical theatres continued to be the site for these collections, as they had been since the 1500s. Many others were found in the homes of anatomists, where the collection, as in the days of Ruysch, served multiple purposes: it allowed the surgeon-anatomist to represent himself as a virtuous and knowledgeable medical practitioner, thus generating both financial and cultural capital for the collector. This was certainly the case for the most famous surgeon-anatomist in the UK: John Hunter (1728–93). His collection eventually became the Hunterian Museum of the Royal College of Surgeons, housed in Lincoln's Inn Fields. It also most certainly applied to the collection of father and son Vrolik and their 'Museum Vrolikianum'.

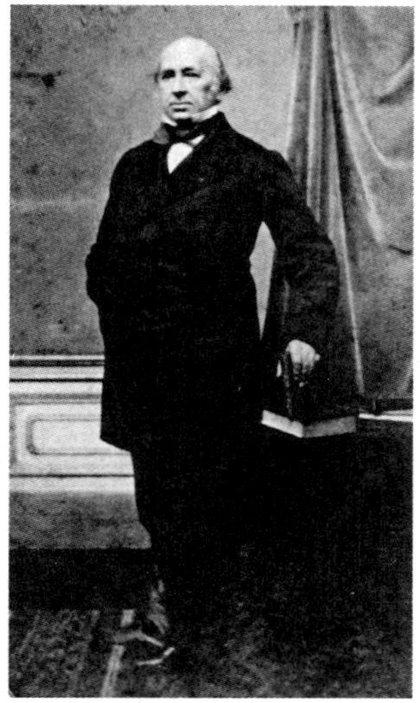

Portrait of Willem Vrolik (1801–63), around 1860

Counting more than 5,000 preparations and specimens, the private collection of Gerard (1775–1859) and Willem (1801–63) Vrolik shared many characteristics of these eighteenth-century anatomical museums, even though it was created during the first six decades of the nineteenth century. It contained examples from a broad range of anatomical study: pathological (morbid) anatomy, comparative anatomy, congenital malformations, foetal development, ethnography, osteology and obstetrics. Located in Gerard's stately mansion on the river Amstel in Amsterdam, it grew into a famous collection that attracted admiring scientists from all over the world.

Gerard, professor of botany, obstetrics, anatomy and surgery, focused his collection on pathological anatomy and obstetrics, while his son, professor of anatomy, physiology and zoology, showed an interest in comparative anatomy and congenital defects. Like some of their famous predecessors, such as Ruysch

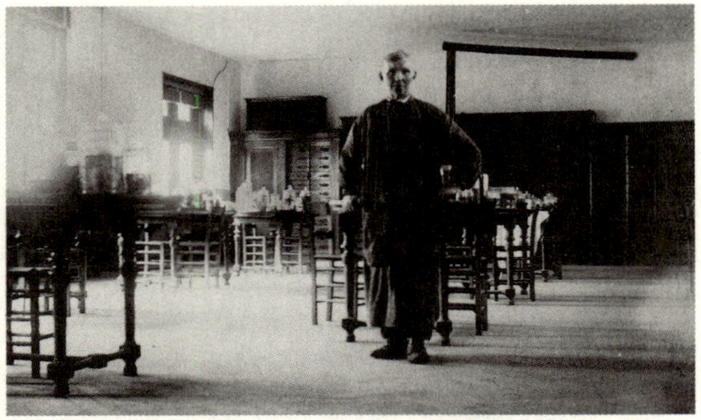

A technician from the Amsterdam anatomical laboratory prepares demonstrations for students in the dissection ward, 1902

and Hunter, father and son Vrolik injected tissue with red wax, dried it or preserved it in alcohol in thin phials. They suspended the specimens with horse hair and sealed the preparation using wax, a pig's bladder and slate.

How did they create such extraordinary specimens? In the days of Ruysch, much of his art was still wrapped in mystery, for instance, he kept the materials and method for injecting his preparations with red wax a secret for most of his life. In the second half of the eighteenth century, the method Gerard Vrolik used *was* known. In 1790, British physician Thomas Pole (1753–1829) offered a practical guide for the creation of anatomical preparations called *The Anatomical Instructor*. To inject a preparation with a red wax-like solution, Pole suggested the following:
'-Yellow bees wax, sixteen ounces
- white resin, eight ounces
- Turpentine varnish, six ounces
- Vermillion, three ounces.'[4]
To create a yellow injection, Pole added 'King's Yellow, two ounces and a half'[5]; for dark blue injections, he put in 'Blue Verditer, 10 and a half ounces'[6].

The anatomical instructor also demonstrates how to create articulated skeletons (mounted skeletons that have the various parts connected in such a way as to demonstrate normal relationships and allow motion between components as in the living body):

> We are first to remove from the bones, the cutis, muscles, tendons, viscera, and every thing except the connecting ligaments, and cartilages, which should be carefully avoided…this being done, lay it in clean water…let it remain until putrefaction has so far advanced as that the soft parts may easily be separated from the bones, when it may be dissected away with a knife, scissors, and forceps, taking care not to injure the ligaments…then remove it into clean lime water, or a solution of pearl-ash for two or three days, to take off any greasiness, and give it a more beautiful white; when it has laid long enough, wash off, with clean water, all the lime that adheres to the surface. When the preparation is thus cleaned and whitened, hang it up to dry, or dry it in a frame, calculated to preserve the posture we wish it to be fixed in.[7]

Father and son Vrolik created many such articulated skeletons, a good number of which still exist. Large skeletons of adult men and women show Pott's disease, rickets and scoliosis. But more refined are the tiny skeletons of foetuses with congenital malformations such as anencephaly, spina bifida and sirenomelia. They are part of a collection of about 600 congenital malformations, that for the most part were brought together by Willem Vrolik and still have scientific and medical relevance today. Like many other anatomists of his day, Vrolik thought that teratology, the study of monsters, was important to understand embryological development. He believed that a specific natural force was necessary to allow a living thing to develop and grow, and he called this force *vormkracht* (formative force). Malformations were important objects to study because they could help Vrolik understand the essential workings of this formative force. In his eyes, for instance, conjoined twins were the result of too much formative force, whereas cyclopes or sirenomelia were the consequence of too little force.

Besides his collection, Willem Vrolik's greatest contribution to the science of teratology was his anatomical atlas on the subject in the 1840s, *Tabulae ad illustrandam embryogenesin hominis et mammalium, tam naturalem quam abnormem*, which won an international prize. When Willem died in 1863, the Vrolik Museum contained 5,103 objects, and his widow was keen to sell them. The collection risked being split up and shipped abroad, but a number of prominent residents of Amsterdam thought it 'a loss for the city' and bought the entire collection, donating it to the Athenaeum Illustre, which became the University of Amsterdam in 1877. The Vrolik Museum was the last big private collection of its kind in the Netherlands, and, over the centuries, most sub-collections have stayed together. It has withstood the test of time and has kept its nineteenth-century soul. It is a time capsule of an era before the big medical and scientific discoveries that changed our world forever.

The nineteenth century witnessed great changes that are still apparent in modern life. It saw the birth of many European nation states such as Germany, Belgium and the Netherlands and the creation of nationalism. It witnessed imperialism and the development of world empires like the British and the French. It created the idea that Europe was the centre of the world. Great inventions were introduced, for example, the steam locomotive and the telegraph that made communication and transportation across countries much faster. The era also saw the invention of photography.

Most importantly for the purposes of this essay, the nineteenth century also witnessed the beginnings of modern medicine: the rise and professionalization of hospitals and methods of safer surgery thanks to the development of antisepsis, anaesthesia and various technical improvements. The birth of scientific medicine includes advances in pathological anatomy, microscopy (and other laboratory sciences), statistics and epidemiology, and, in their footsteps, the hygienist movement and public health. These all provided a new empirical and scientific basis for the previous worn practices of surgeons and vague medical theories.

Pathological anatomy proved to be very important for anatomical museums. Developed in Paris in overcrowded hospitals such as the Hôtel-Dieu, pathological anatomy was explored by physicians like Guillaume Dupuytren (1777–1835), Pierre Louis (1787–1872), Xavier Bichat (1771–1802) and René Laennec (1781–1826, the inventor of the stethoscope, a symbol of modern medicine). By taking careful note of patients' accounts of their medical history and by studying their physical state using such objective methods as the stethoscope, the Parisian clinicians did their best to accurately form a diagnosis. When a patient died, even after surgery or other treatment, a systematic autopsy was performed and a report made. Thus symptoms and epidemiology were linked to post-mortem findings. Pathological anatomy, therefore, gave diseases an anatomical (and later histological and cellular) basis. So important were these advances that pathological anatomy and the methods used in the Paris hospitals were introduced all over the Western world. In the UK, for instance, physicians Thomas Hodgkin (1798–1866) and Robert Carswell (1793–1857) introduced the system to London, while John Thompson worked this way in Edinburgh. Thomas

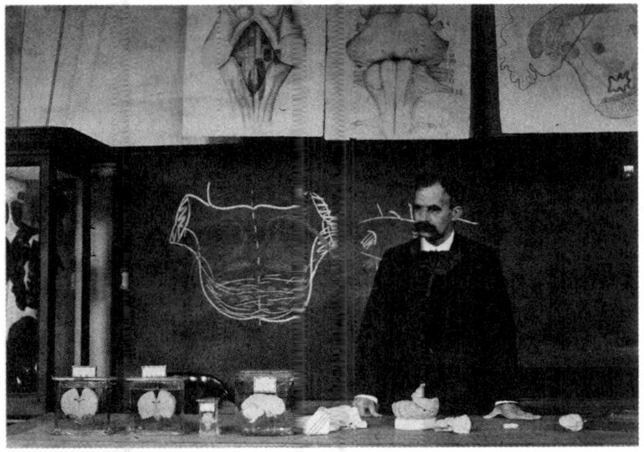

Anatomist Louis Bolk (1866–1930) uses museum specimens to illustrate his lecture on neuroanatomy, around 1918

Dent Mütter (1811–59, after whom Philadelphia's anatomical Mütter Museum is named) also studied in Paris in the early 1840s.

Pathological anatomical preparations have always formed part of anatomical collections, but from the middle of the nineteenth century they were systematically collected and became the focus of most university, hospital-based anatomy museums and were used by doctors and students as study aids. A 'typical' body had been seen plenty during dissections and surgeries; it was the 'abnormal', the morbid, the pathological, that carried great interest. The collections became libraries or reference points that were always expanding.

Most of the nineteenth-century anatomical museums and collections that still exist are pathological anatomical museums. Well-known examples are the Musée Dupuytren in Paris, Barts Pathology Museum and the Gordon Museum of Pathology in London, the Mütter Museum of the College of Physicians in Philadelphia, the Berlin Museum of Medical History at the Charité and the enormous Federal Pathalogical Anatomical Museum in Vienna.

While pathological anatomical collections grew over the course of the nineteenth century, one of the most famous pathologists of his time, German Rudolph Virchow (1821–1902), advocated a different approach. He studied disease to the cellular, microscopic level. To him, pathology was not just a macroscopic empirical support in the identification of a disease or a tool to clinical medicine as a whole, but rather a science of its own, a laboratory science like (bio)chemistry and, most notably, experimental physiology. These brought what was known as the laboratory revolution in medicine. And physiology, which had been under the umbrella of anatomy, became a separate discipline with its own chair and laboratory.

Due to advances in physiology and pathological anatomy, normal (gross or systematic) anatomy was no longer considered to be the cornerstone of medicine. It was still accepted as an essential component of medical education, but it no longer defined it.

In the Netherlands, pathological anatomy and physiology grew distinct from normal anatomy over the course of the 1850s and 1860s. From that point onwards, new professors for the chair of anatomy were hard to find. In Amsterdam, the medical faculty decided to appoint a German anatomist. This choice was not surprising: Germany's scientific approach to medicine had outpaced France as the leading country. It proved to be a lucky decision for Amsterdam as it revived interest in anatomy and its importance not only for education, but also for scientific progress. This revival also led to new anatomical collections that focused on healthy anatomy instead of morbid anatomy.

In late-nineteenth-century Germany, as in the Netherlands, anatomy was taught in specialized laboratories or institutes that were linked to universities. Students of medicine followed anatomy lectures, which mainly dealt with systematic anatomy, the study of the structure of such organ systems as the skeleton, the muscles, the nervous system, the blood vessels or the internal organs. They also undertook a practical aspect of anatomy in the form of dissections, which focused on the gross or topographical side of anatomy: the specific locations of nerves, organs, muscles and bones within the body. Over the course of a few weeks, students dissected such things as the head and neck, the brain, the thorax, the abdomen and the limbs. The aim was that by combining their theoretical knowledge with their practical experience, students gained a deep understanding of the anatomy of the human body.

In Germany there were several 'schools' of anatomy. Many of these focused only on the practical application of anatomy for future doctors. Some, however, added a more fundamental scientific purpose to the act of dissecting and studying anatomy. One such anatomist was Karl Gegenbaur (1826–1903), who along with zoologist Ernst Haeckel (1834–1919), was one of the greatest advocates of Charles Darwin's theory of evolution in Germany. They developed a research field called evolutionary morphology, through which they tried to unravel the evolutionary tree by studying the form (morphology) of all kinds of animals. As for human anatomy, their focus lay on the crucial moments in human evolution that could be traced to distant or close relatives of man.

Two of Gegenbaur's pupils, Max Fürbringer (1846–1920) and Georg Ruge (1852–1919), were appointed in succession as professors in Amsterdam. Both created anatomical preparations of animals for the Vrolik Museum. Ruge's pupil and successor, Dutch anatomist Louis Bolk (1866–1930), continued Ruge's work, acquiring over 4,000 preparations, of which half were animal specimens, including over 800 skulls of monkeys and apes.

Fürbringer, Ruge and Bolk also collected anatomical abnormalities, not so much the more severely malformed foetuses, like Gerard and Willem Vrolik had amassed earlier, but instead more subtle anatomical variations that they discovered with their students during dissections. Their discoveries showed that people lived with these variations unknowingly and without any harm being caused. Examples included extra ribs or vertebrae (see p.139), perforated sternums (see p.133), lungs with an extra lobe or horseshoe kidneys (see p.103). For students and doctors, knowledge of these variations was important. They showed the medical world that no one is the same on the inside. As such, they served a comparative purpose as specimens of morbid anatomy in pathology museums.

The greater part of their collection comprised 'normal' anatomy of both the adult and the developing foetus. Bolk, in particular, was dedicated to collecting examples of normal human development. In the early 1900s, he appealed to other doctors in the country to send him aborted foetuses and stillborn

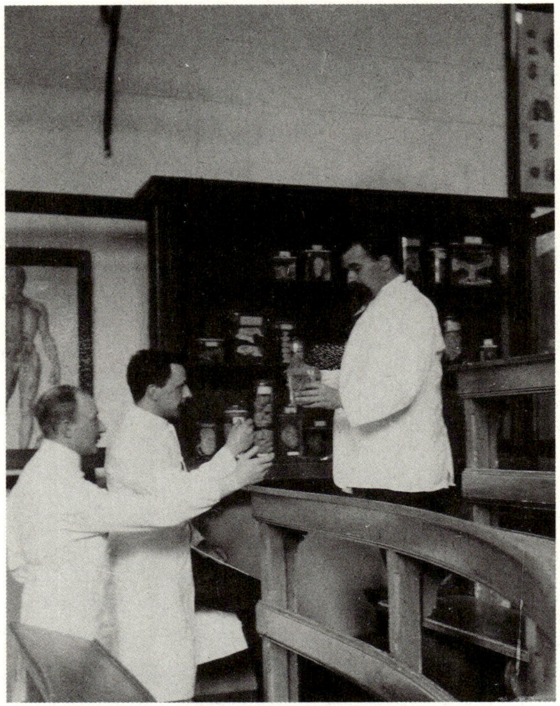

Louis Bolk and his assistants handle anatomical preparations in the lecture hall of the Amsterdam anatomical laboratory, 1905

infants in the *Nederlandsch Tijdschrift voor geneeskunde* (*Dutch Journal of Medicine*). With these, he created a highly instructive series on foetal, brain and heart development, which still forms an important and unique part of the Vrolik.

Fürbringer, Ruge and Bolk and their students also created a large number of preparations showing the topographical and regional anatomy of the normal adult. Ironically, these 'normal' specimens make the museum very special. Most anatomical museums do not feature normal adult anatomy because it was not considered useful (gross anatomy was practised during dissections). Moreover, when the supply of corpses was low and the demand for other types of anatomical demonstrations grew – which was the case in the Netherlands just as it was in the UK – large, ready-made demonstration preparations had their limitations. Big, hand-blown glass jars were very expensive, as was 60+% alcohol, especially in the quantities needed for a leg, trunk or head. Furthermore, the jars were round and distorted their contents, making it necessary to remove the specimens in order to study them. This resulted in spilt alcohol, broken glass and lids that did not fit well. Alcohol also evaporated over time. Anatomical models made from wax or plaster, dry skeletal specimens and dried preparations, therefore, were used much more often to teach normal anatomy.

Two inventions from the last decades of the nineteenth century changed this. The first was the invention of formalin; the second, the creation of square glass vessels. Formalin reduced costs greatly because preparations only needed to comprise about 4% formaldehyde, while the remainder of the solution was water. Neither did it evaporate, although it did acidify over time. In Germany, the glass industry invented new methods to create square jars, the advantage being that the specimen was not distorted by the shape of the glass and did not need to be taken out of the jar to be studied.

These two inventions were crucial in the development of Fürbringer and Ruge's collection of normal anatomy in Amsterdam. The creation of a museum collection in large square jars would still have been a costly endeavour, but luckily for the Amsterdam anatomical laboratory, the administrators of the University of Amsterdam were amenable. Thus quite a large number of excellent normal systematic and topographical preparations were created. The dissections were carried out by the professor himself or by skilled assistants and students. It is these preparations, along with the foetal and embryological specimens and the congenital malformations, that make the Vrolik Museum remarkable. These preparations were not put in jars to simply create a record, but rather as a means to understand human

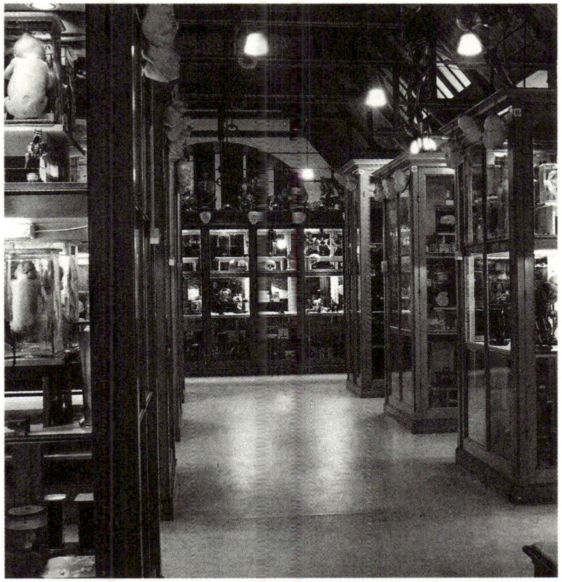

Museum Vrolik, the Amsterdam anatomical museum, early 1980s

anatomy. Today it is not just medical students who benefit from these specimens, but a growing public use them as a guide to their own bodies, as readers of this book will acknowledge. The collections added by Fürbringer, Ruge and Bolk more than doubled the size of the Vrolik Museum. Thanks to them, the museum became the biggest and most complete museum of (normal) human anatomy in the Netherlands.

In 1984, the Vrolik moved from Bolk's anatomical laboratory of 1909 to the newly built Academic Medical Center of Amsterdam, a huge building combining medical care, teaching and research. The AMC remains the home of the Vrolik Museum, but in 2012, after almost 30 years, the museum space was thoroughly renovated and the permanent exhibition radically changed. Over 2,000 anatomical preparations are now on display, including the Vrolik's congenital malformations and morbidly deformed adult skeletons and trunks; Fürbringer's collection of animal morphology; Ruge's topographical and regional anatomical preparations; and Bolk's series of foetal development.

As this book proves, we remain fascinated by the structure of the body and how it functions, its complex and intricate systems. Yet so much of human anatomy remains a mystery, and we have much to learn as we advance along the crooked path carved out by the history of medicine.

Laurens de Rooy is a medical historian and the Curator-director of the Vrolik Museum in Amsterdam.

Notes

1 See Simmons, John E., *Fluid Preservation: A Comprehensive Reference*, Rowman & Littlefield, 2014.
2 Alberti, Samuel J. M. M., *Morbid Curiosities: Medical Museums in Nineteenth-Century Britain*, Oxford University Press, 2007, p.14.
3 Ibid., p.15.
4 Pole, Thomas, *The Anatomical Instructor*, Couchman and Fry, 1790, p.21.
5 Ibid., p.22.
6 Ibid., p.23.
7 Ibid., pp.152–53.

A Few Words About Stereoscopy

DENIS PELLERIN

Everybody has heard of 3D, and millions of movie-goers the world over have experienced 3D films. Most people, however, think that 3D arrived with James Cameron's *Avatar* (2009), few realizing that stereoscopy – as 3D was known at the time – is a nineteenth-century invention even older than photography since it was officially presented to the world in 1838, before Frenchman Louis-Jacques-Mandé Daguerre (1787–1851) and Englishman William Henry Fox Talbot (1800–77) demonstrated their respective photographic processes, the former on a silver-coated copper plate (daguerreotype, 1839) and the latter on silver-coated paper (calotype or talbotype, 1840).

For centuries, great minds tried to understand the principles of binocular vision and to find irrefutable evidence as to why and how our two eyes give us a unique three-dimensional image of the world. Some, like Leonardo da Vinci (1452–1519), came close to describing the process, but it was not until 1838 that British physicist and polymath Charles Wheatstone (1802–75)[1] designed a simple apparatus that showed beyond doubt that our two eyes give slightly different pictures of our surroundings and that the brain uses these small discrepancies to produce a single three-dimensional image that enables us to perceive the depth of objects and compute their distance from us.

Wheatstone focused on this age-old problem some time towards the end of 1832, but as he was also working on the electric telegraph at the time it was not until June 1838 that this Professor of Experimental Philosophy at King's College published a paper entitled 'Contributions to the Physiology of Vision. – Part the First. On some Remarkable, and Hitherto Unobserved, Phenomena of Binocular Vision'. With this paper was an instrument Wheatstone had named a stereoscope (from two Greek words, *stereo* meaning 'solid' and *scopos* meaning 'see'). It consisted of two mirrors set at a right angle, on either side of which stood an easel holding the monocular picture of a simple object seen respectively from the left and right eyes. When viewers pressed their noses to the part where the two mirrors joined, each of their eyes could only see the flat picture that was meant for it, but the two images coalesced into a three-dimensional one that had the appearance of a solid object. Stereoscopy was born.

In order for the brain not to be distracted or assisted in any way by depth cues such as shadows, Wheatstone made use of basic outline figures such as a cube, a cone and the frustum of a square pyramid. He wrote:

> For the purposes of illustration, I have employed only outline figures, for had either shading or colouring been introduced it might be supposed that the effect was wholly or in part due to these circumstances, whereas by leaving them out of consideration no room is left to doubt that the entire effect of relief is owing to the simultaneous perception of the two monocular projections, one on each retina.[2]

Although Wheatstone's article and research were lauded by his peers, his instrument was too cumbersome to be used outside a physics laboratory and had no commercial value owing to the lack of a proper way of obtaining with enough accuracy the two pictures necessary to create the illusion of depth. As soon as photography was revealed to a stunned world, Wheatstone was quick to see the advantages it could bring to his invention and soon approached the inventor of the calotype and the first French and English daguerreotypists, asking them to produce images for his instrument. The resulting pictures, sadly, have disappeared, but from what Wheatstone wrote about the calotypes produced by Talbot and Henry Collen (1797–1879), they were far from satisfactory and would probably not be recognized as stereoscopic pictures today. As for the daguerreotypes obtained, it is almost certain that their shiny mirror-like surface would not have proved easy to fuse in an instrument using mirrors.

The stereoscope, therefore, stayed in the laboratory for several years until Sir David Brewster (1781–1868) designed a more compact and more portable instrument, using prisms instead of mirrors, which he presented to the Scottish Society in 1849. Brewster's lenticular or prismatic stereoscope was no bigger than a pair of binoculars and allowed the two flat pictures of the stereoscopic pair to be mounted side by side. Brewster had optician George Lowden from Dundee build a prototype, but the two men fell out and Brewster could not find anybody else interested in building his apparatus. In the spring of 1850, Brewster and his daughter made their way to Paris where they met with the Abbé Moigno (1804–84), a Catholic priest, lecturer and expositor of science. Moigno saw the potential of the stereoscope and took Brewster to the shop of opticians Jules Duboscq (1817–86) and François Soleil (1798–1878), who agreed to make the instrument and to bring it to the Great Exhibition that was to be held in London the following year.

According to a story craftily created by Brewster himself, and for which there is no other corroboration than his own words, Queen Victoria saw the stereoscope on Duboscq's stand during one of her thirty or so visits to the exhibition. She was allegedly so impressed that her keen interest made it popular almost overnight. Duboscq was inundated with orders and to thank the sovereign made a luxurious stereoscope that was presented to her, on his behalf, by Brewster.

None of the above can be proved, and perhaps the truth is a little less romantic. The stereoscope was one of the 100,000 or so objects displayed inside Crystal Palace and it passed so unnoticed that there is no mention of it in the press until two days before the exhibition closed. What attracted the Queen's and Prince Albert's attention was not Duboscq's stand, but the stereoscopic pictures of Crystal Palace made by photographer Antoine Claudet (1797–1867) who, for the next couple of years, was the main advocate and maker of stereoscopic pictures in the UK. He was gradually joined by Richard Beard (1801–85), William Edward Kilburn (1818–91), Thomas Richard Williams (1824–71) and a few others, all of whom specialized in binocular portraits for the daguerreotype. The pictures that have survived are probably among the best examples of early stereoscopic imagery and most of them are masterfully tinted. As they were rare and quite expensive, the photographs were generally carefully preserved in leather cases or wooden boxes and have remained extraordinarily fresh, sometimes as pristine as when they were made.

After 1851, thanks to the invention of the wet-plate collodion process – a new photographic technique on glass by British photographer Frederick Scott Archer (1813–57) – photography, and more particularly stereo photography, was to know an unprecedented boom that was further encouraged by the development of albumen paper. By 1853, it was possible to obtain economical but very sharp glass negatives that could be made into as many paper prints as desired. Stereoscopic pictures, being quite small (around 18 × 9cm when mounted), needed shorter exposure times than larger formats and could be used in all kinds of lighting conditions. This quickly encouraged photographers to leave their studios and take their photographic equipment into the streets, then the countryside and even far-away countries. At the same time, the stereoscope inched its way into the minds of the people thanks to its introduction, often alongside the microscope, in fêtes, conversazioni, soirées, lectures and scholarly societies.

In 1854, the London Stereoscopic Company was created and soon became the largest publisher of stereoscopic pictures in the UK. After some careful planning of its distribution circuit through retail agents all over the country, the company rapidly expanded its collection of stereoscopic slides and boasted 10,000 titles as early as March 1856. That year, it published three catalogues covering a variety of subjects, among which were statuary, miscellaneous subjects of the 'Wilkie' character,[3] views of the Crystal Palace by T. R. Williams, 'scenes in our village' also by Williams, views of the great Crumlin Viaduct in Wales, groups from Shakespeare's *The Winter's Tale*, and a collection of views of Pompeii and Naples.

In the second half of 1856, David Brewster published his book *The Stereoscope: Its History, Theory, and Construction*,[4] in which he not only explains binocular vision and the principles behind the stereoscope, but also devotes several chapters to the various uses of the instrument. It appears that this book gave rise to the stereoscopic craze since photographers soon made sure that nearly every aspect approached by Brewster became a stereoscopic reality in France and the UK.

The year 1857 marks the start of what was quickly dubbed 'stereoscopomania' on both sides of the English Channel. The first stereos were copyrighted, in Paris and in London, and many advertisements began to appear in the press. French photographic magazine *La Lumière* inaugurated a column devoted exclusively to the stereoscope and its various features. Collections of stereoscopic slides were reviewed in the specialist and general press and, in the UK, the London Stereoscopic Company launched its bold campaign 'No home without a stereoscope'. At the same time, dashing entrepreneurs, quickly realizing that it would become impossible for anyone interested in stereoscopic pictures to buy them all, created lending libraries from which people could rent a dozen cards and a stereoscope for an evening or a few days. More ardent fans could borrow as many cards as they wished for a period of time.

It is difficult to imagine nowadays the revelation experienced by the average person, who was suddenly able to see the world in three dimensions, without leaving the fireside, and to travel in space without any discomforts. Up until this point, newspapers and magazines only published engravings or lithographs, and a person's only true representations of the world were those they had seen with their own two eyes. By the end of 1859, most European countries and the USA, India, China and Japan could be visited through the oculars of the magical instrument. Other subjects included portraits of celebrities (Charles Dickens in 1857, Napoleon III and his arch-enemy Victor Hugo in 1858, dancer Lola Montez in 1859, to name but a few), hundreds of sentimental or comic scenes staged in photographers' studios, historical tableaux (Henry Tudor on the battlefield at Bosworth, the death of Thomas Becket, the capture of Joan of Arc, Mary, Queen of Scots, signing her abdication and escaping from her prison at Loch Leven), scenes copied from or inspired by famous paintings or illustrations,[5] the most picturesque streets and monuments of main cities in the UK, France, Germany, Italy, Holland and Spain, and striking examples of the ingenuity of man (Isambard Kingdom Brunel's Great Eastern steamship and Clifton Suspension Bridge, and the tubular Britannia Bridge across the Menai Strait, Anglesey, by Robert Stephenson and Francis Thompson).

Born and developed in the UK and France, stereoscopic photography arrived in the USA in the late 1850s and was soon popularized by Oliver Wendell Holmes (1809–94) after he created an economical viewer and published several articles on the subject in *The Atlantic Monthly*. American views flooded the European market from the late 1880s onwards and started a second golden age for the stereoscope. In its heyday, Underwood & Underwood, a major stereographic company in the USA, was producing over 25,000 images per day.

What about human anatomy? In around 1857, Cleer Sewell Alger (1819–92), a photographer from Diss, Norfolk, published at least three binocular photographs of a human skeleton: a front view, a back view and a close-up of the head and torso. They have no caption and are very much at odds with the rest of his output, making it difficult to know whether they were made at a customer's request or as part of his ordinary stock. One thing is certain: they were meant as an aid to anatomical study and certainly would have proved cheaper to buy (and store) than a full skeleton. In 1858, a gruesome arrangement of six human skeletons around a table was published under the title *The Skeletons' Carouse, or the Bon(e) Vivants*. At the time, it was considered to be in very bad taste and attracted strong disapproval and many harsh comments in the press. A journalist from *Photographic News* wrote in the 15 October 1858 issue of the magazine:

> We cannot find words strong enough to express our disapprobation of the publication of this slide. There is something about the whole affair so flippantly sacrilegious, that it cannot fail to disgust any right-thinking person. It is well-known that, even when science demands that a body should be subject to medical examination, there is always a strong feeling against such a proposal; and it is often only by urgent representations that the relatives of deceased persons can be induced to allow the bodies of their dead friends to be thus mutilated. But what must we say of this gross violation of all the laws of decency and propriety? Were not all these six skeletons at one time living men, who moved and breathed, and took part in the duties of life as we do now?

The journalist concluded: It is to our minds, the result of a wanton profanity, which would turn into ridicule what ought to be held in religious veneration', condemning the whole thing as 'shameless irreverence'.

With such strong words and such strong feelings towards the human body, even after death, there is little wonder that stereographic photographs of human anatomy from this period are, unlike series on natural history, quite rare. There are a few odd cards but no sets as such, and it was not until the beginning of the twentieth century that such stereoscopic publications as *The Edinburgh Stereoscopic Atlas of Anatomy* by D. J. Cunningham and David Waterston (published between 1905 and 1911, 10 volumes) *The Stereoscopic Skin Clinic* by S. I. Rainforth (1910, with 132 colour stereoscope plates) and *Stereoscopic Studies: Pathological Anatomy of the Eye* by Anton Elschnig (1900, with 64 stereograms) appeared. All were destined for the medical profession, not the general public, and were intended as visual aids to the study of anatomy with no artistic aspirations.

After World War II, however, stereo photography was far from being as popular as it had once been. Nevertheless, new series of anatomical pictures were released, among them a *Stereoscopic Atlas of Neuro-Anatomy* by H. D. Rubinstein and C. L. Davis (1947, with 43 plates) and, more surprisingly, the seventeen-year project *A Stereoscopic Atlas of Human Anatomy*, by David L. Bassett and William B. Gruber (1962, 23 volumes). The latter consisted of 221 View-Master reels of 7 pictures each, and a shorter version was republished recently with 83 of the original 1,554 stereo views.

As can be seen from this very brief survey of stereoscopic photography, Jim Naughten's work, though unique in many ways, belongs to a long tradition that started in the early days of the stereoscope and has continued, with a series of ups and downs, to the present day. It is time to let Naughten's pictures speak for themselves and to travel in space and time (some of the pieces are nearly two centuries old) through the oculars of what was called 'the optical wonder of the age': the stereoscope. In the isolation created by the magical instrument, we are not only gazing at art, but also at ourselves as a species, as a part of the animal world that still holds so many secrets and so many promises.

Denis Pellerin is a photographic historian and a member of the London Stereoscopic Company

Part of this essay has been published previously in the book *Animal Kingdom*.

Notes

1 Wheatstone was also interested in musical instruments and was the inventor of the concertina.

2 Wheatstone's paper was published in the scientific journal of the Royal Society of London *Philosophical Transactions*, vol. 128, pp. 371–94.

3 David Wilkie (1785–1841) was a Scottish artist famous in his lifetime for his genre paintings and historical subjects.

4 Sir David Brewster, *The Stereoscope: Its History, Theory, and Construction*, London, John Murray, 1856.

5 An exhibition devoted to this 'Poor Man's Picture Gallery' took place at Tate Britain between October 2014 and October 2015, and showed side by side the original paintings and the stereoscopic cards they had inspired, or, in a few cases, had been inspired by.

Foetuses

SIX-MONTH-OLD FOETUS SHOWING BONE DEVELOPMENT
Twentieth century

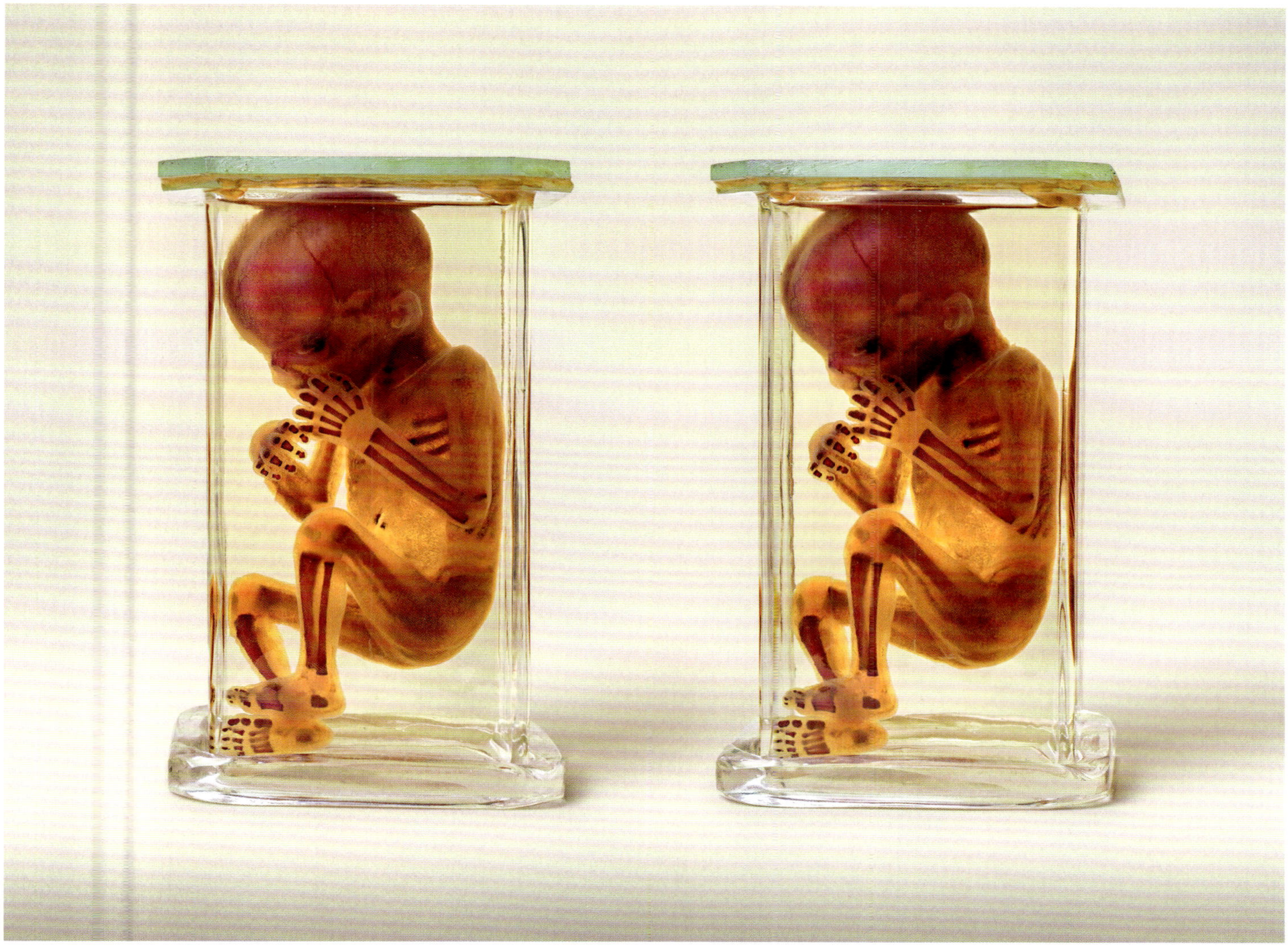

SKELETON OF A FOETUS WITH SIRENOMELIA
Before 1850

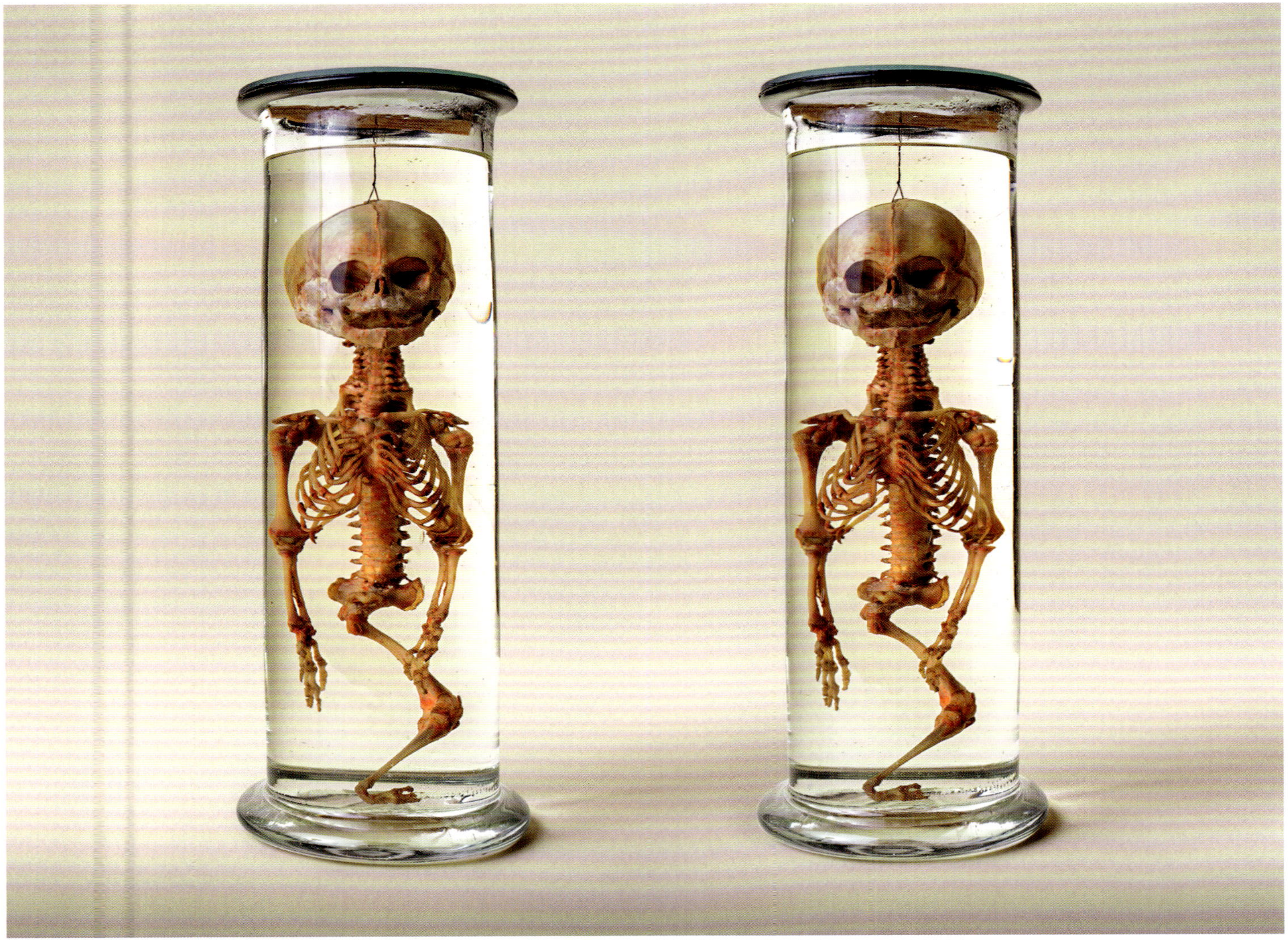

SIX-MONTH-OLD FOETUS IN UTERUS
1898

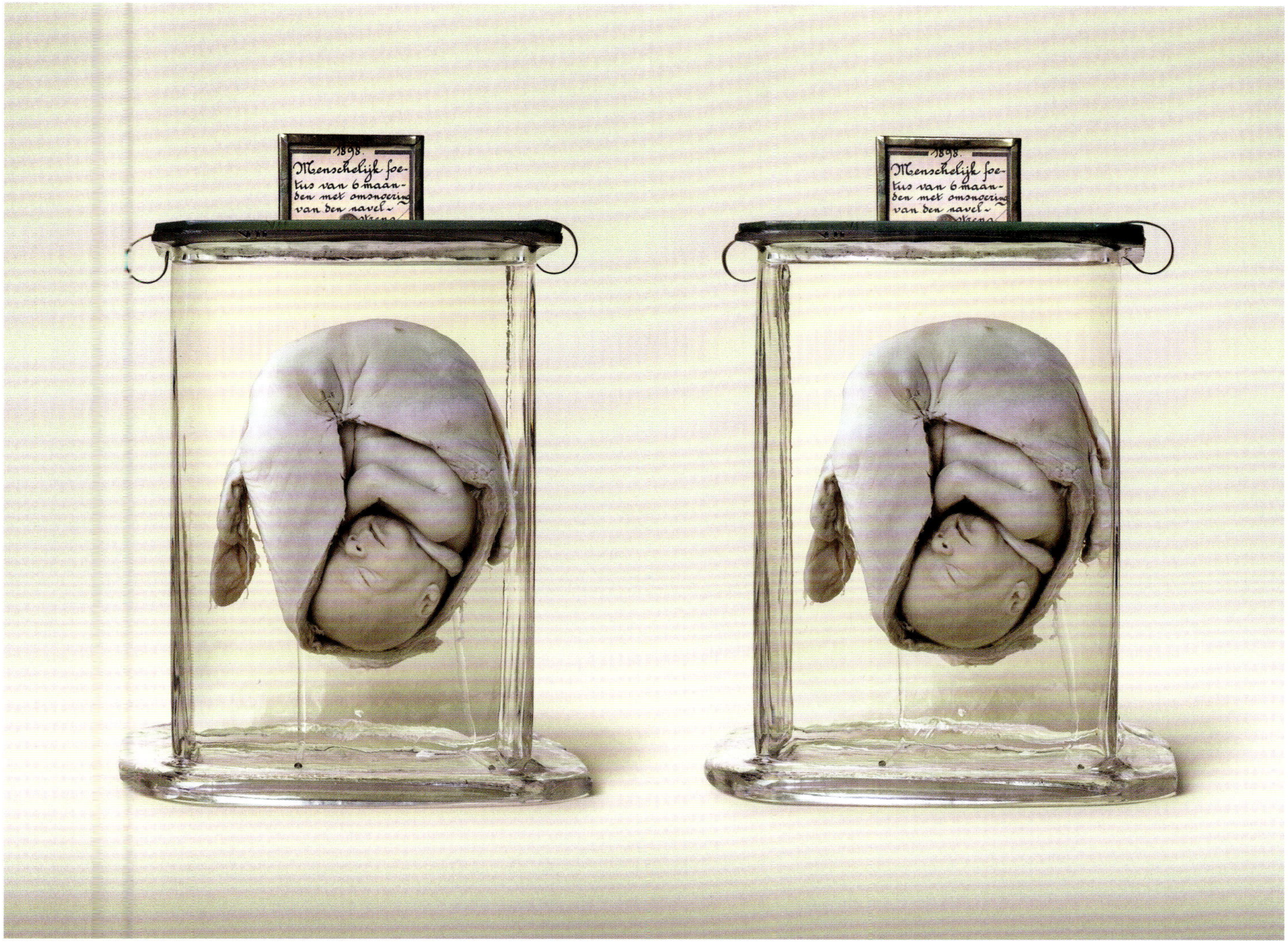
1898.
Menschelijk foe-
tus van 6 maan-
den met omsnoering
van den navel-

TRIPLETS

1937

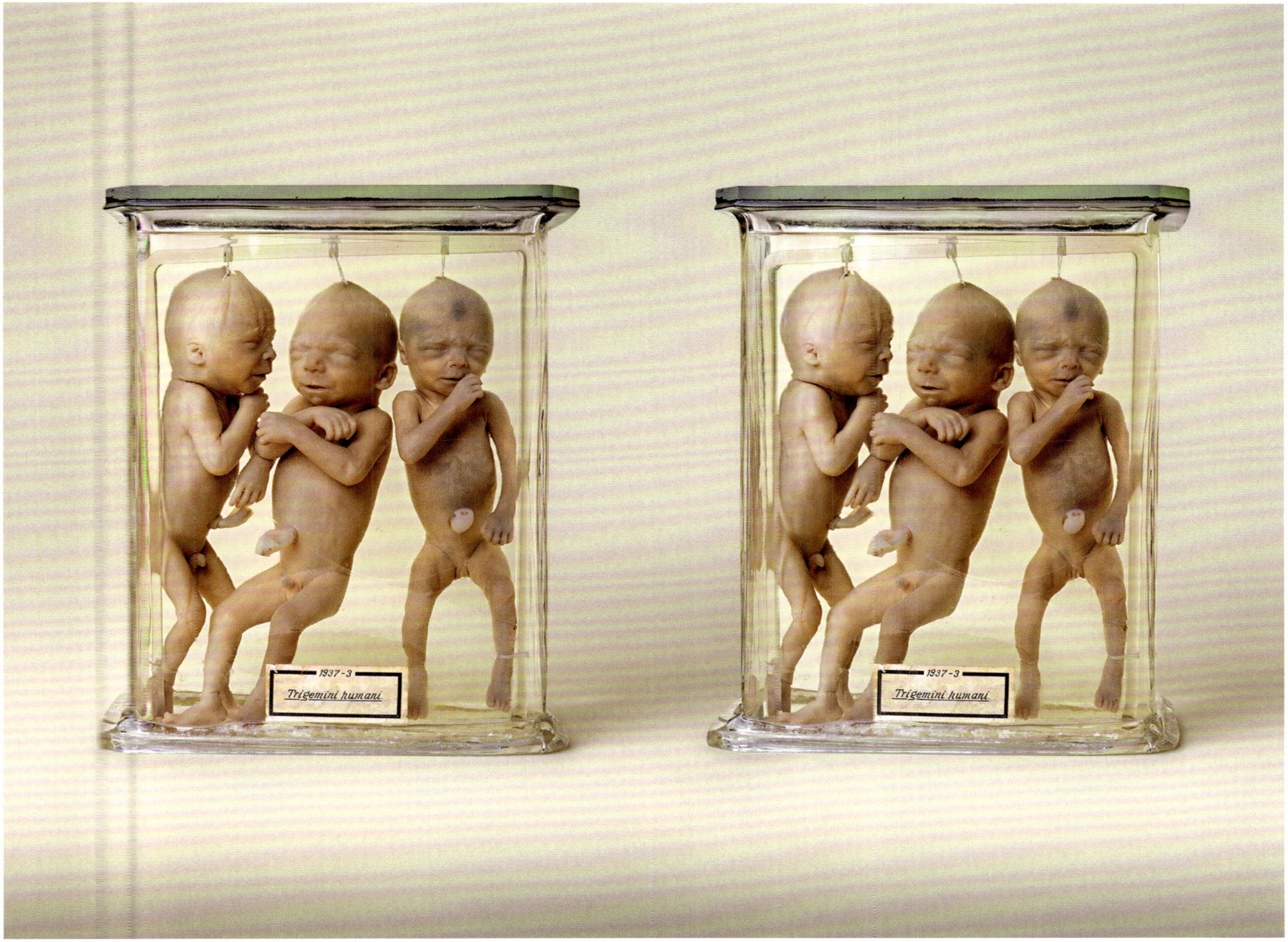
1937-3
Trigemini humani
1937-3
Trigemini humani

HEAD AND BRAIN OF AN EIGHT-MONTH-OLD FOETUS

1902

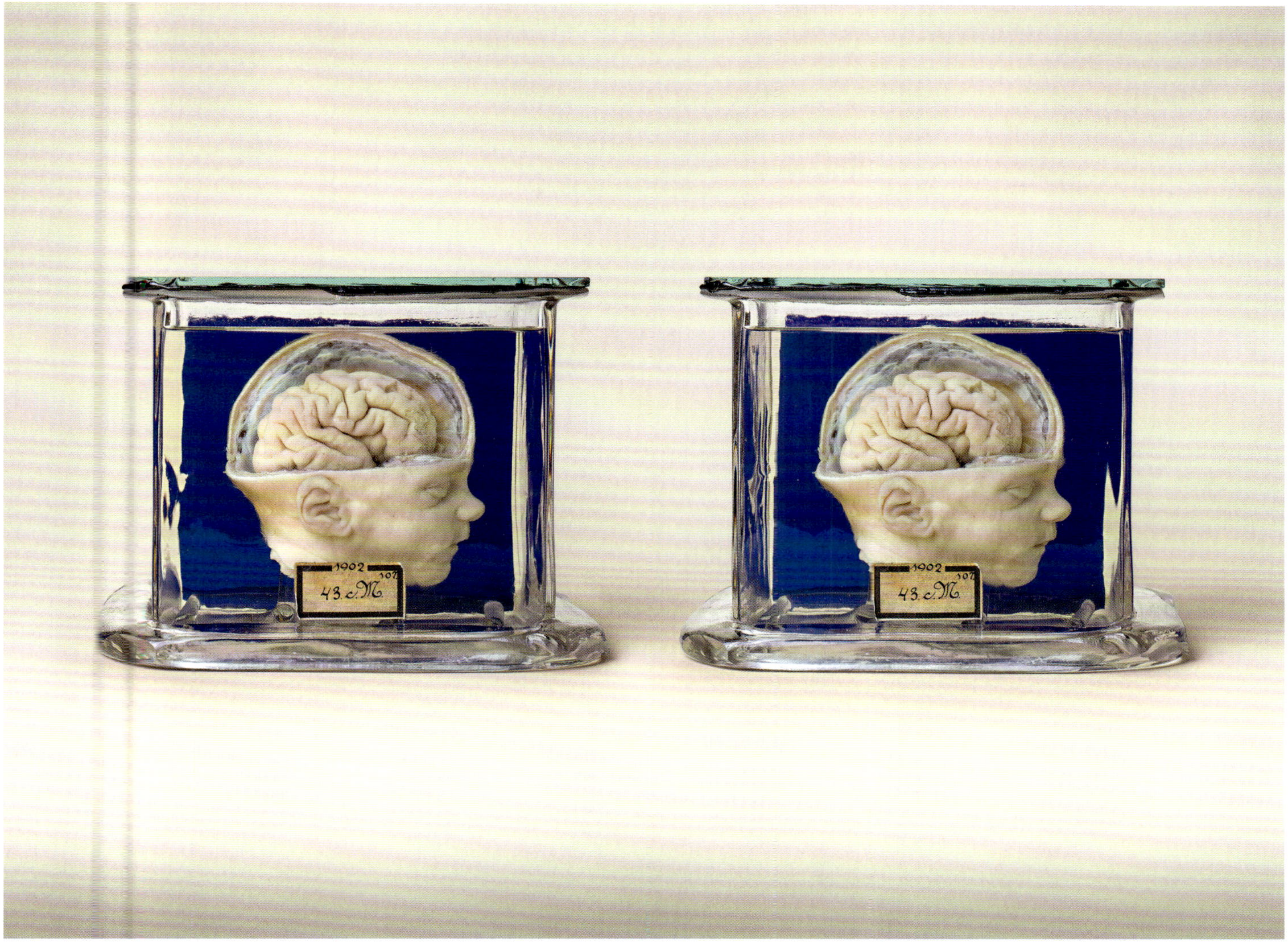
1902
107
43. c. M.
1902
107
43. c. M.

CONJOINED TWINS

Between 1850 and 1880

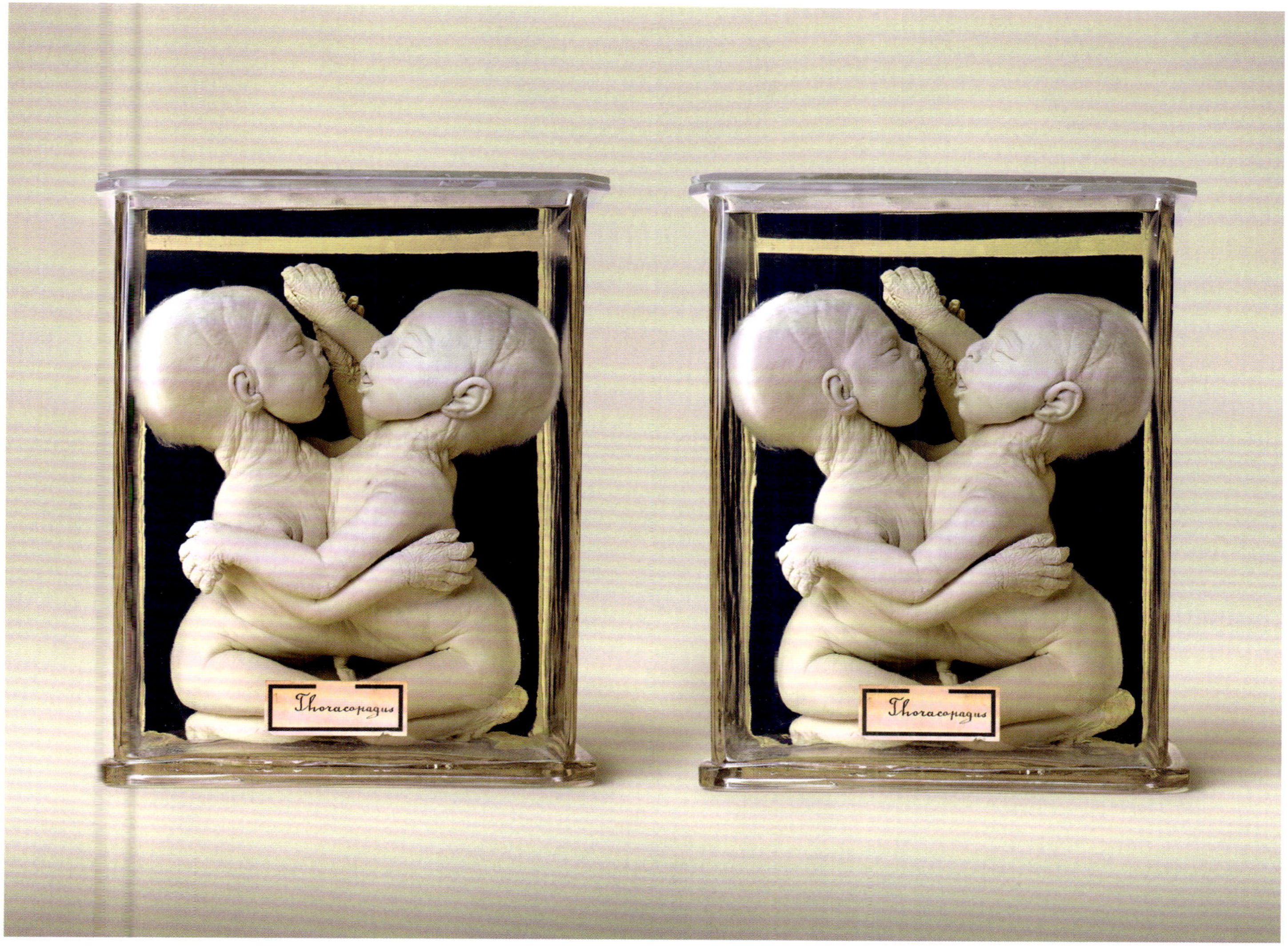
Thoracopagus
Thoracopagus

FOETUS SHOWING BLOOD VESSELS

1939

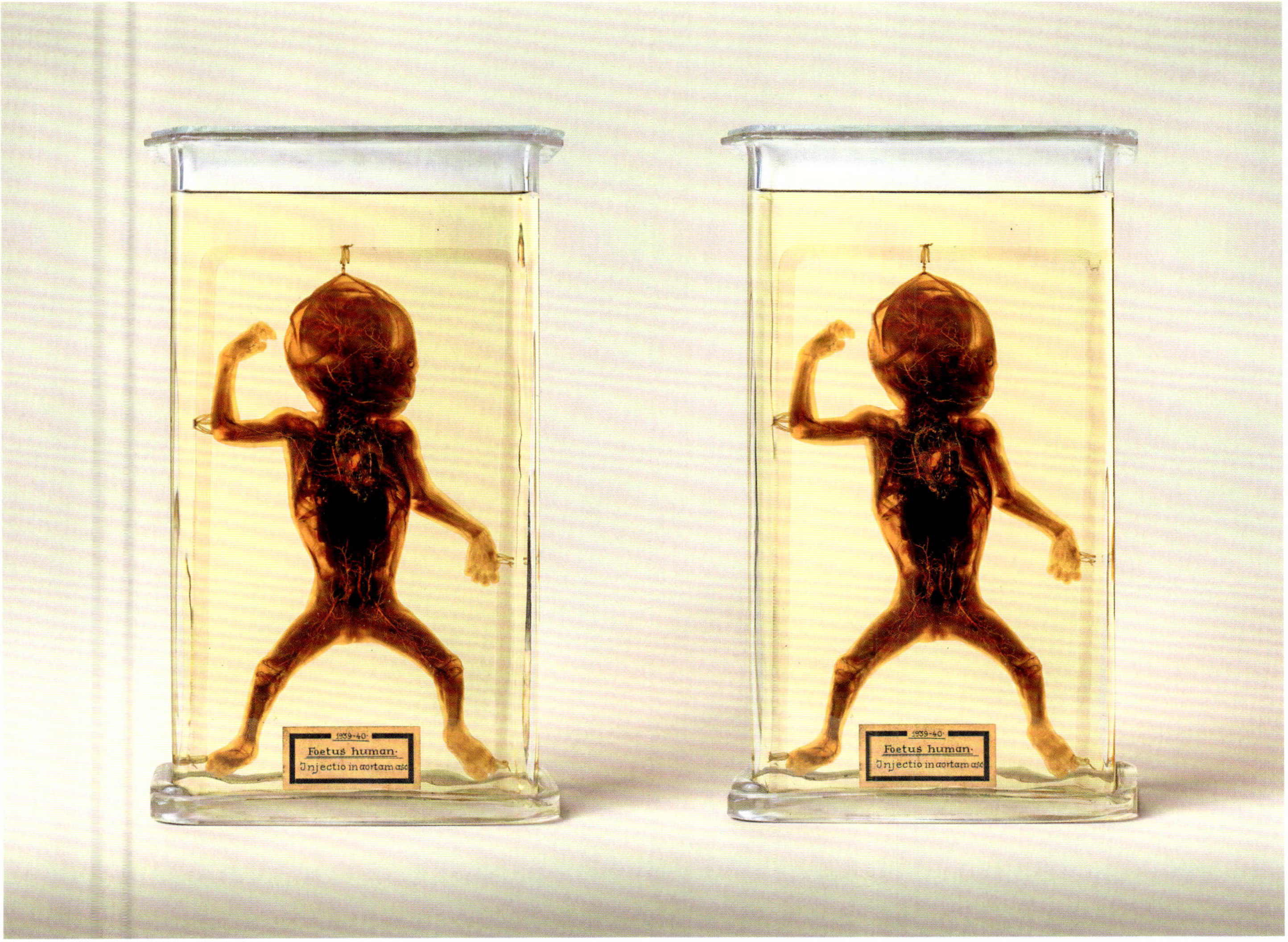
1939-40·
Foetus human·
Injectio in aortam asc
1939-40·
Foetus human·
Injectio in aortam asc

SKELETON OF A THREE-MONTH-OLD FOETUS
Between 1900 and 1930

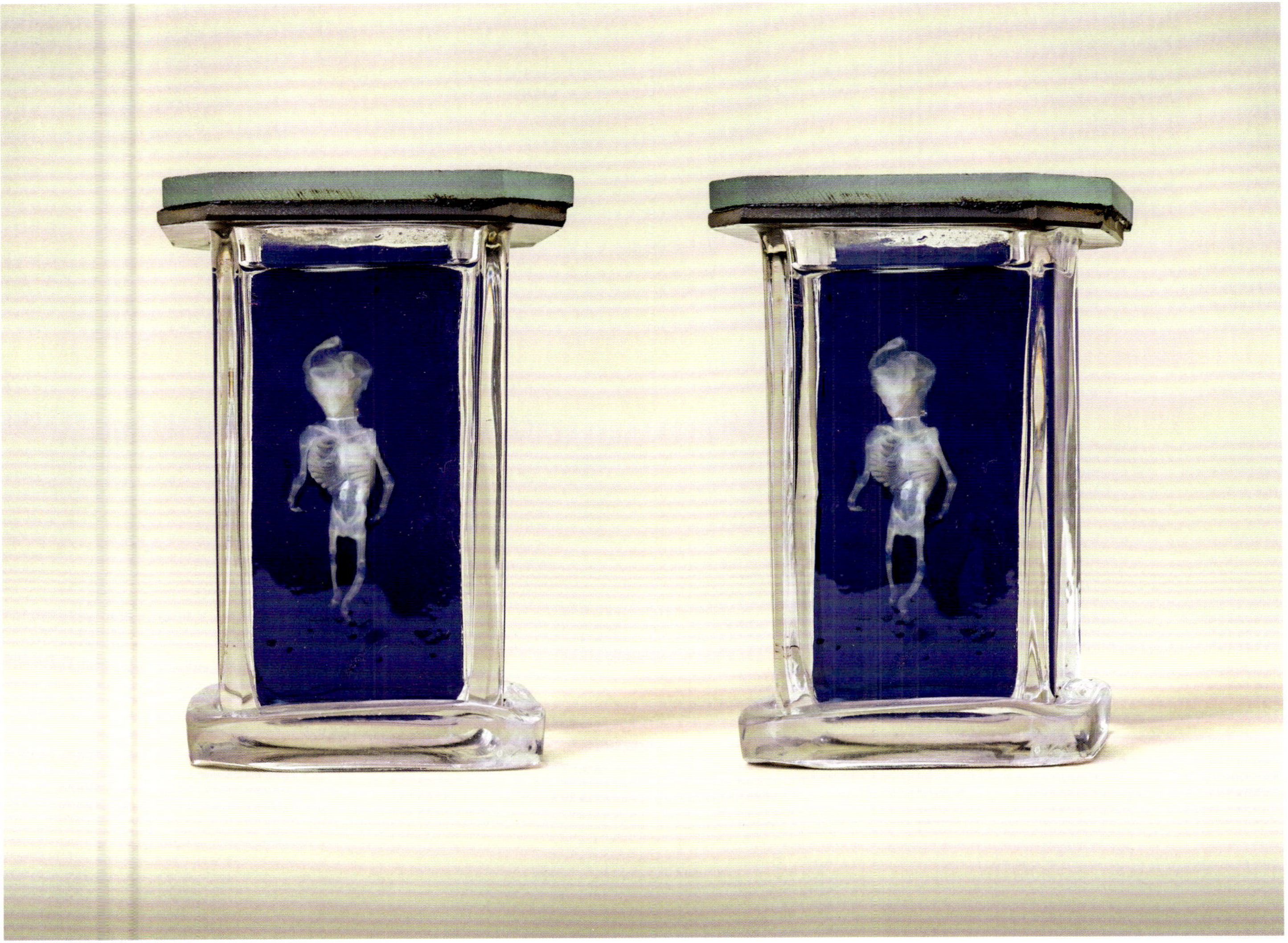

CYCLOPS
Before 1836 (relabelled in 1926)

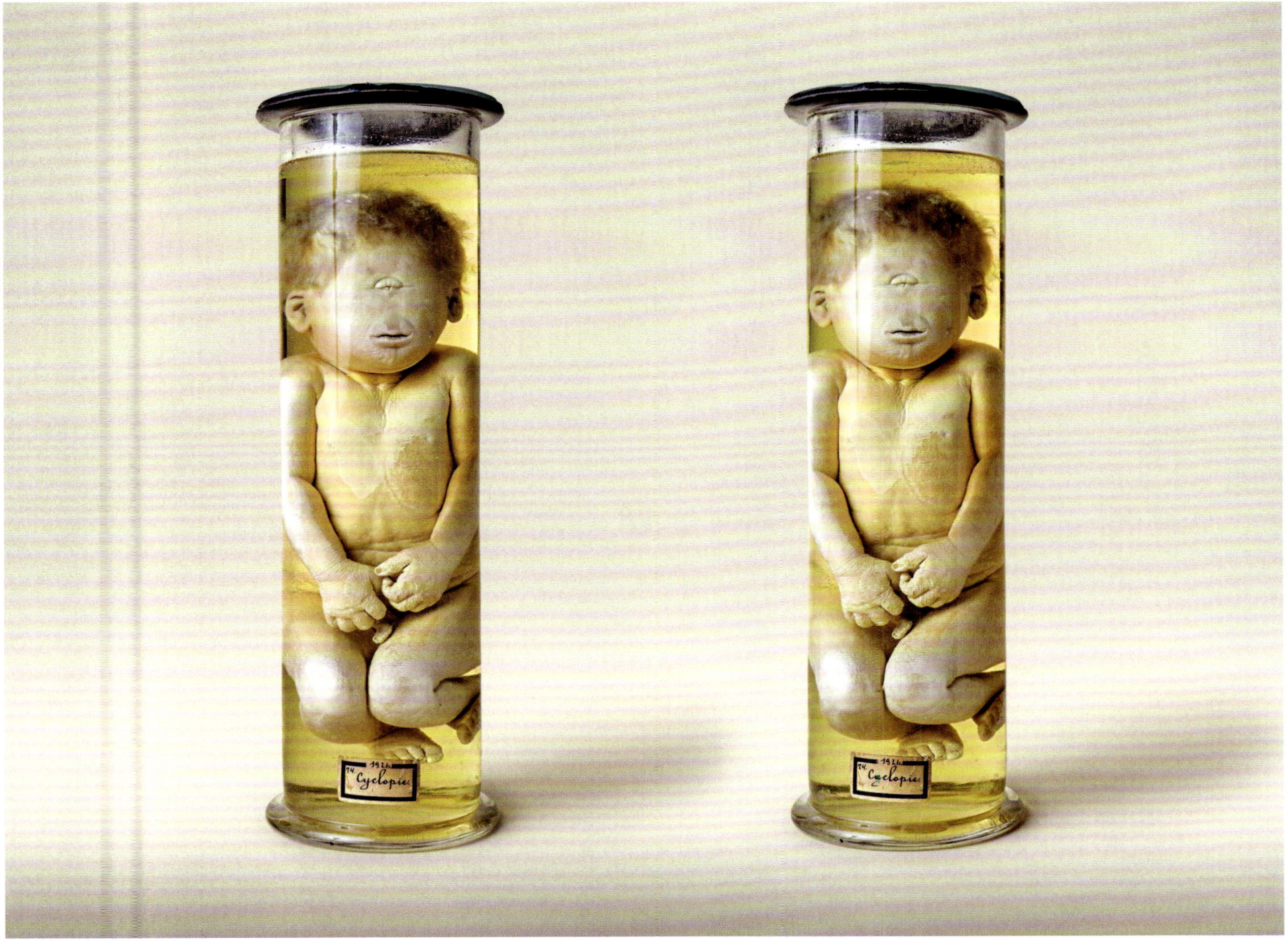
Cyclopie
Cyclopie

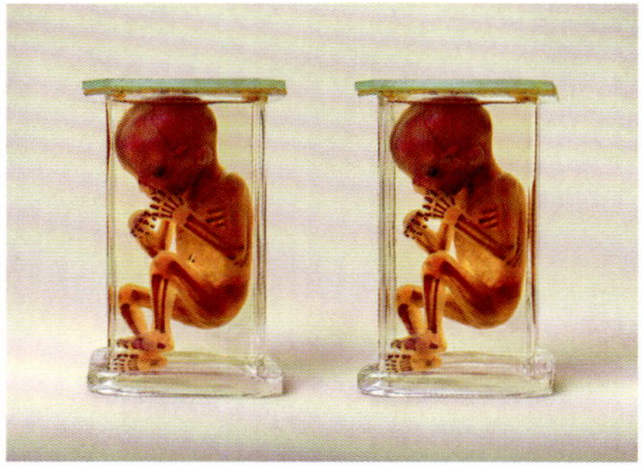

Six-month-old foetus showing bone development

Twentieth century

The conserving fluid makes the tissue translucent, while the skeleton has been coloured with an orange dye that attaches to bony tissue and shows the development of ossification in the foetus. A large part of the bones has already ossified, but the process is far from finished: the joints of the knee, elbow, hand and foot still consist largely of cartilage, which has not been dyed by the colorant.

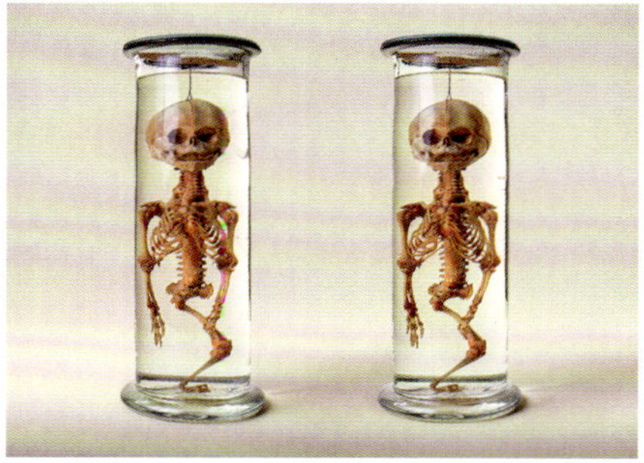

Skeleton of a foetus with sirenomelia

Before 1850

Sirenomelia is a rare congenital malformation in which the legs fuse into one lower limb. The condition is also referred to as 'mermaid syndrome', and is named after the sirens in Greek mythology. The severity of sirenomelia varies. Here the common limb consists of one femur, one tibia, a very rudimentary fibula and four toes. The position of the limb has changed: the hip joint is located at the back of the pelvis and the knee bends to the front instead of the back.

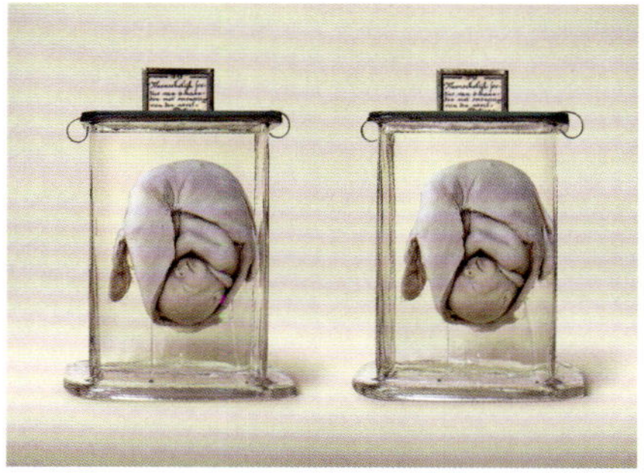

Six-month-old foetus in uterus

1898

Positioned with its head down, the foetus, which is around 24 to 27 weeks, is in the position usually adopted from the 36th week. At six months, foetuses ought to be turning around constantly in the uterus. The umbilical cord is wrapped around the foetus's neck, but it poses no real threat at this stage (it is only during delivery that it may be a problem).

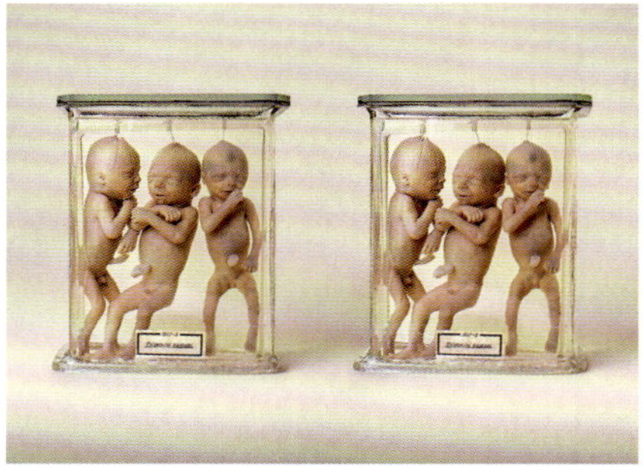

Triplets

1937

The triplet originated from a dizygotic (fraternal) twin in which one of the two zygotes split and developed into a monozygotic (identical) twin. Twins and triplets tend to be born several weeks earlier than a single infant. In this case, it is most likely that the medical care of the 1930s was not capable of keeping the girl and two boys alive.

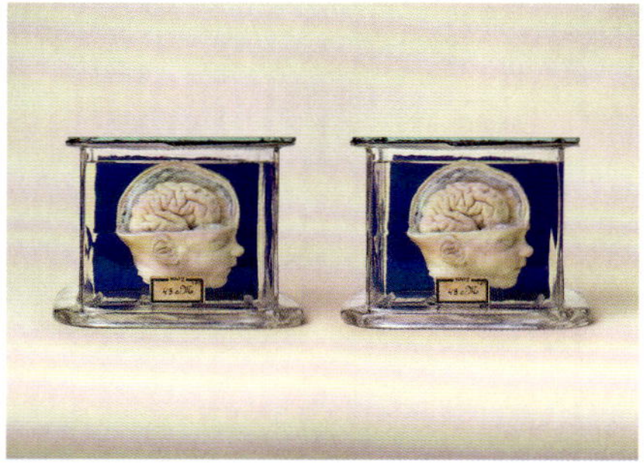

Head and brain of an eight-month-old foetus

1902

The cranial bones have been partially removed to show the brain of the developing foetus (43cm). The dura mater is visible between the brain and the skull. A considerable space exists between the brain and dura to allow for the fast growth of the brain during development

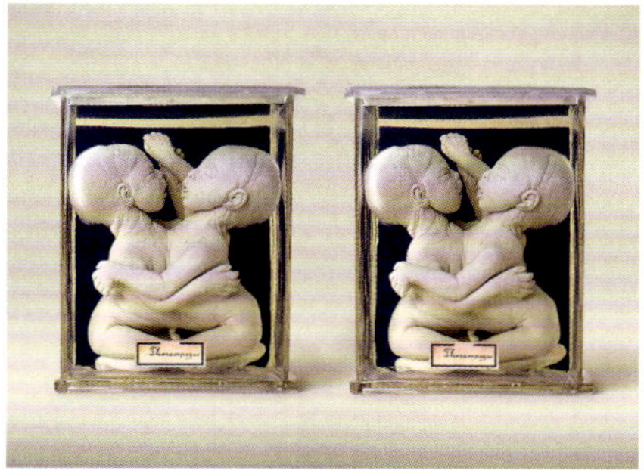

Conjoined twins

Between 1850 and 1880

The term Siamese twins derives from famous conjoined twin brothers Chang and Eng (1811–74), who came from Siam, now Thailand. They were not the first Siamese twins, but they gained notoriety because they travelled the world during the nineteenth century. There are several types of conjoined twins, although the join is always at the same point on both bodies: for example, chest to chest, side to side or head to head. The majority of conjoined twins – around 75% – are joined at the chest and/or abdomen, as in this case.

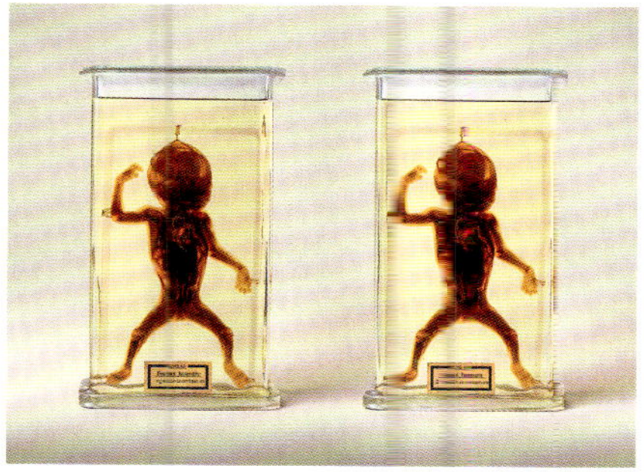

Foetus showing blood vessels

1939

Conserved in a clearing agent, a fluid that makes the tissue translucent, the arteries of the foetus have been injected with an orange-red dye. The large blood vessels in the lumbar region and arms are clearly visible. Moreover, the ramification of these larger vessels into many smaller ones can be seen in the limbs, the trunk and especially in the head. It shows how the circulatory system delivers blood to the leg muscles, brain and other organs.

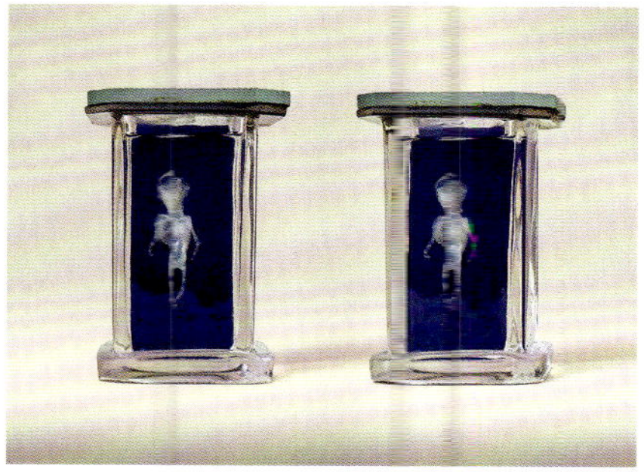

Skeleton of a three-month-old foetus

Between 1900 and 1930

At first, the foetal skeleton consists mainly of cartilage, but during development, both before and after birth, the cartilage is gradually replaced by bony tissue. This process begins in the foetus's third month of growth and takes place in the ossification centres of the ribs and the long bones of the limbs.

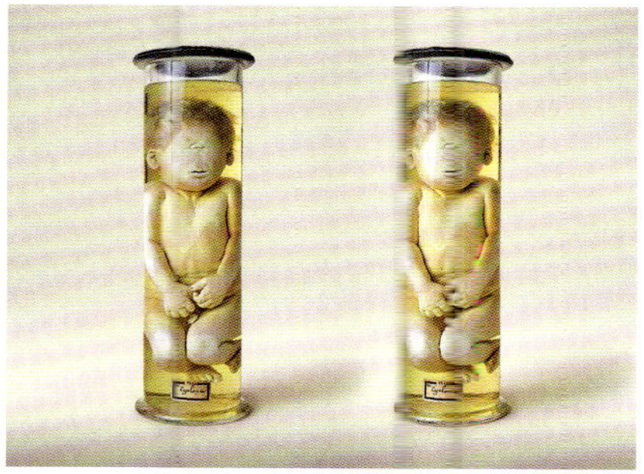

Cyclops

Before 1836 (relabelled in 1926)

Cyclopia results from an abnormal embryological development of the brain in which the lateralization (the development of the brain to the sides) fails to take place properly. Thus structures of the brain that normally would develop to the side, originate centrally, whereas more central parts of the brain do not develop at all. Since the development of the eyes correlates to the development of the brain and the development of the face follows both these, the eyes are not located to the left and right of the nose. Instead, just one fused eyeball forms where the nose should be.

Nervous System & Sensory Organs

EYE AND EYEBROW
Before 1830

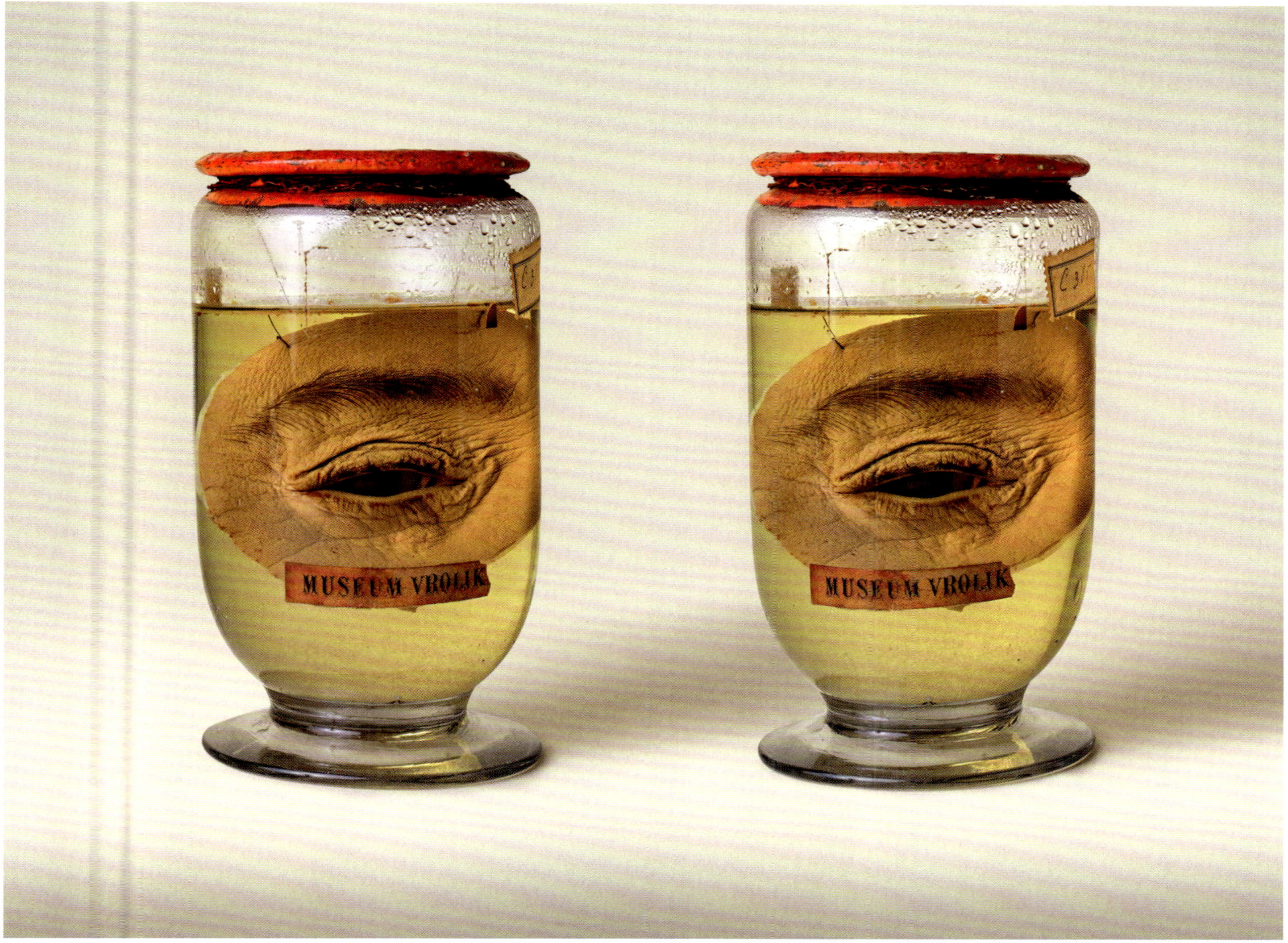
MUSEUM VROLIK
MUSEUM VROLIK

HEAD AND BRAIN OF AN ADULT MALE
1898

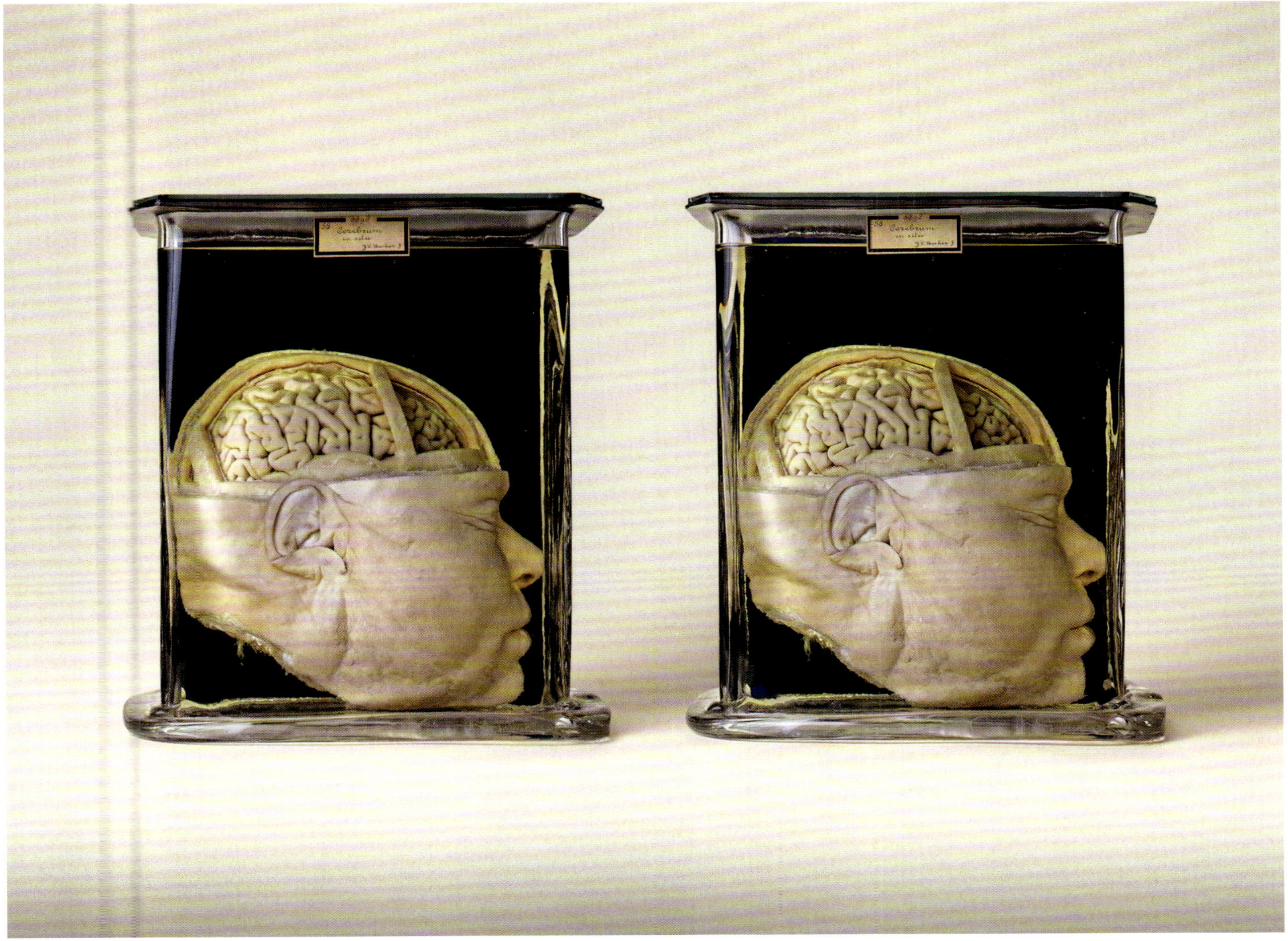

TWO CROSS SECTIONS OF THE FACIAL SKELETON
1886

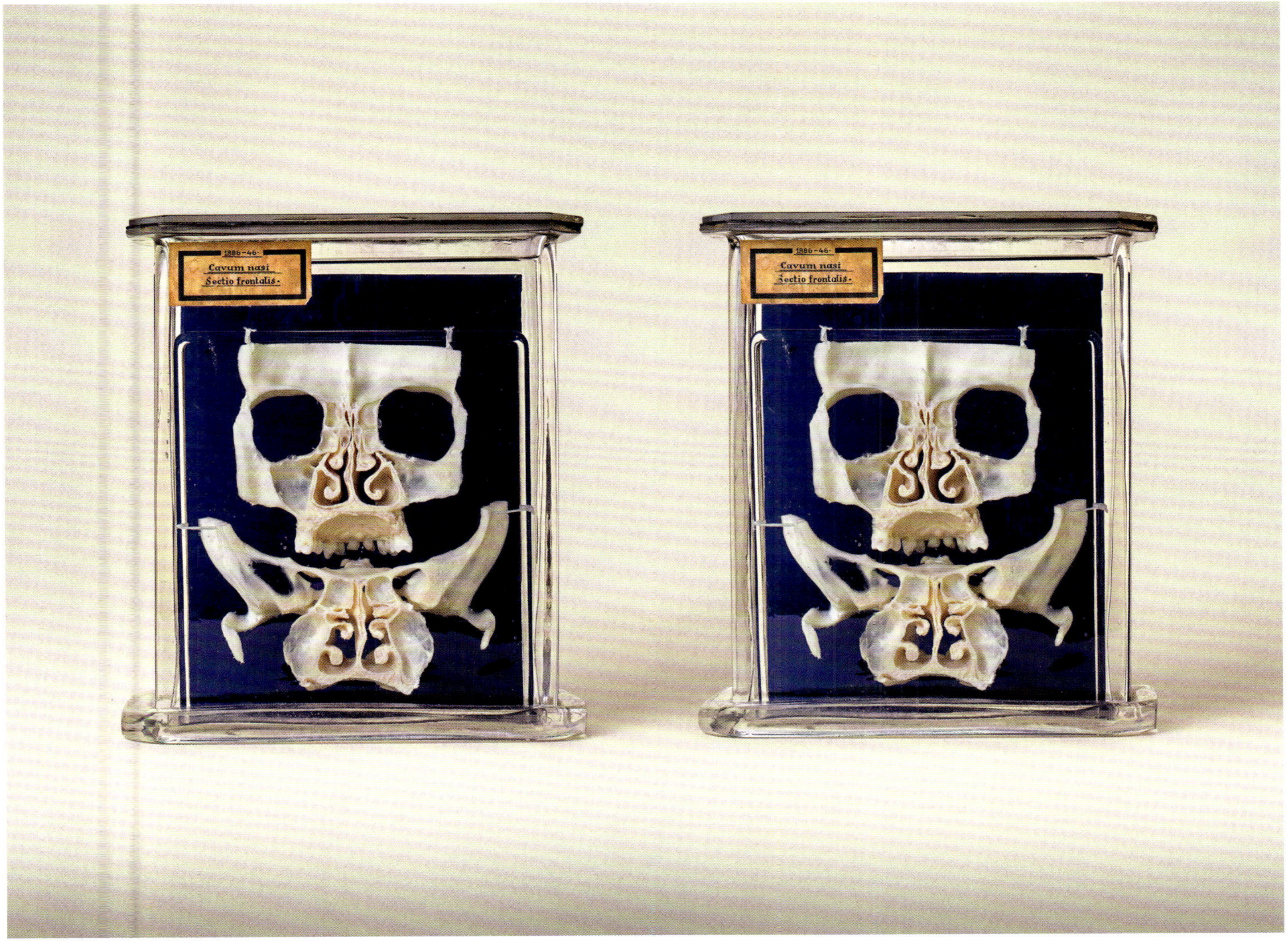
1886-46-
Cavum nasi
Sectio frontalis.
1886-46-
Cavum nasi
Sectio frontalis.

HEAD SHOWING NERVES OF THE CHEEK
1899

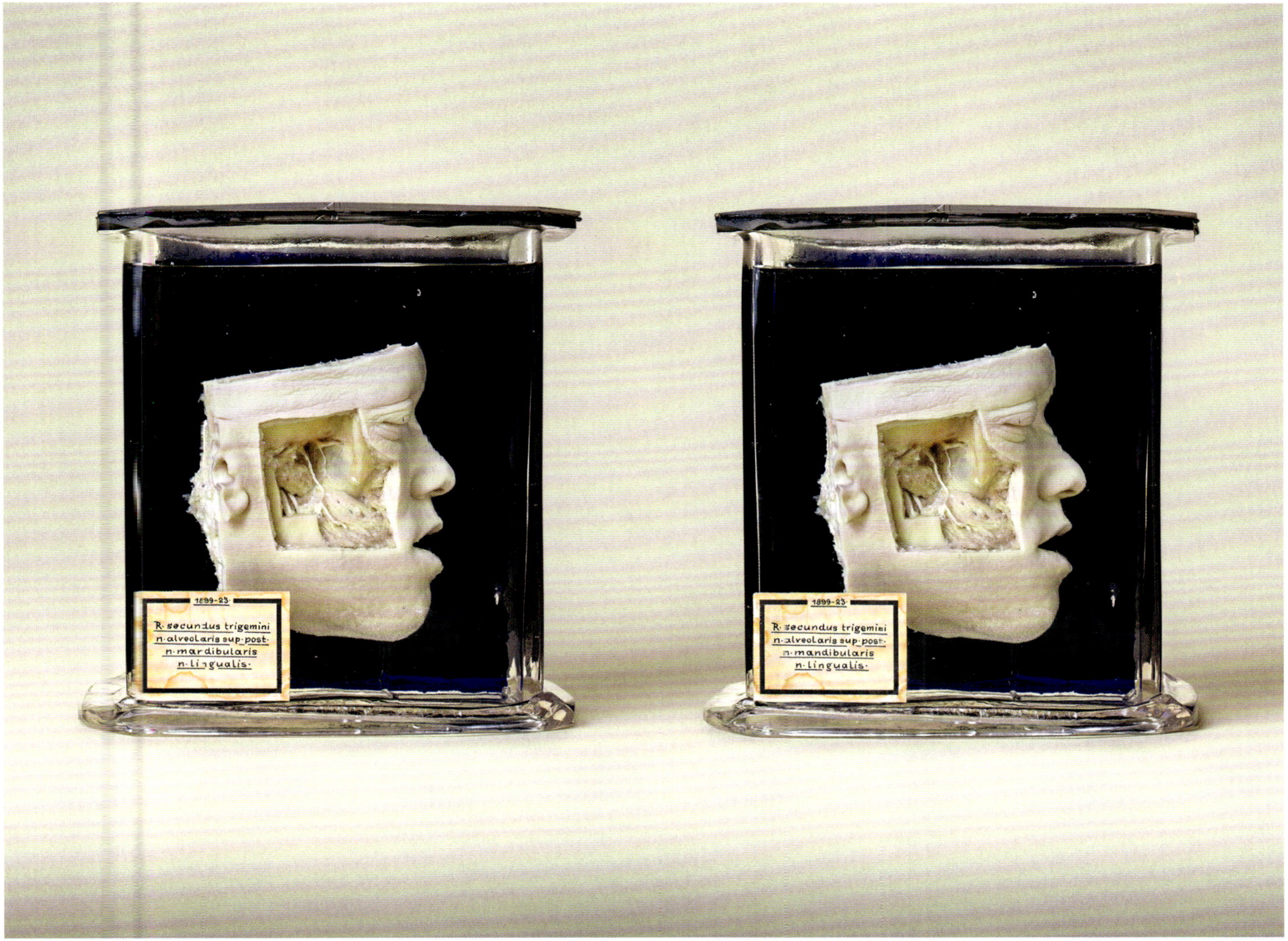
1899-23·
R· secundus trigemini
n·alveolaris sup·post·
n·mandibularis
n·lingualis·

SPINAL CORD OF AN INFANT
1899

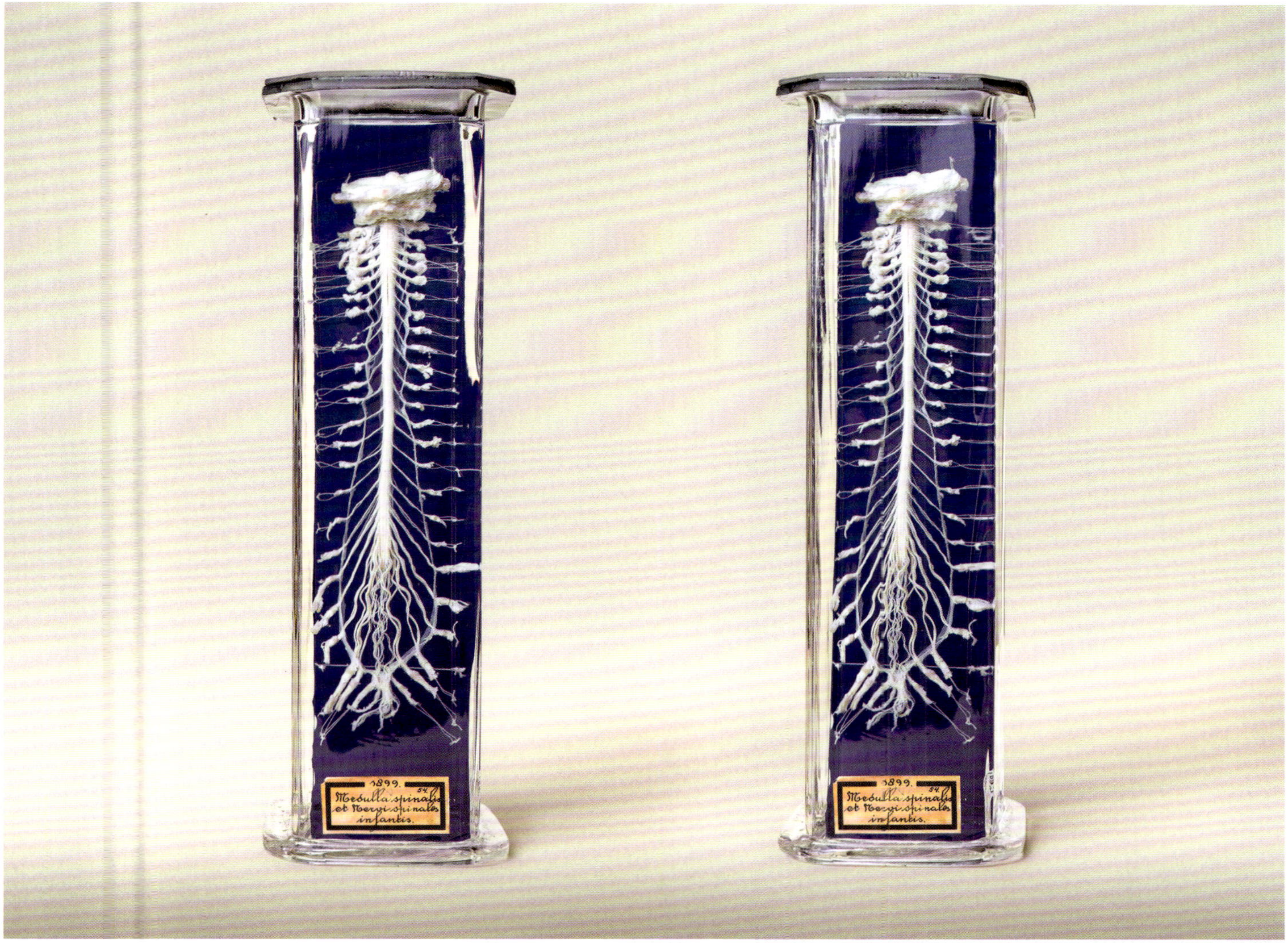
1899.
54.
Medulla spinalis
et Nervi spinales
infantis.
1899.
54.
Medulla spinalis
et Nervi spinales
infantis.

SUPERFICIAL NERVES OF THE HEAD
Between 1863 and 1877

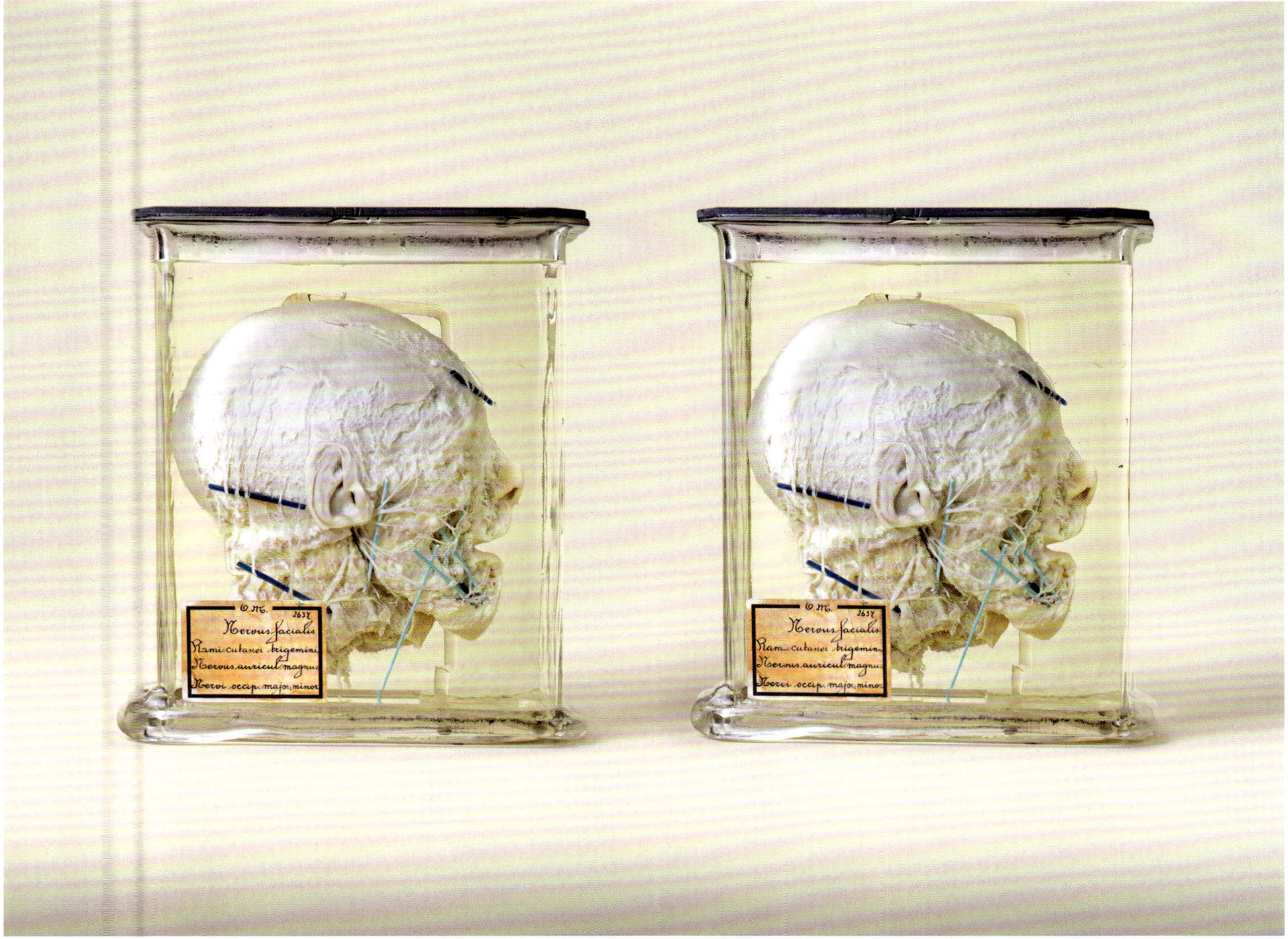
O.M. 2637
Nervus facialis
Rami cutanei trigemini
Nervus auricul. magnus
Nervi occip. major, minor

LEFT SIDE OF ADULT HEAD SHOWING NERVES OF THE CHEEK
1899

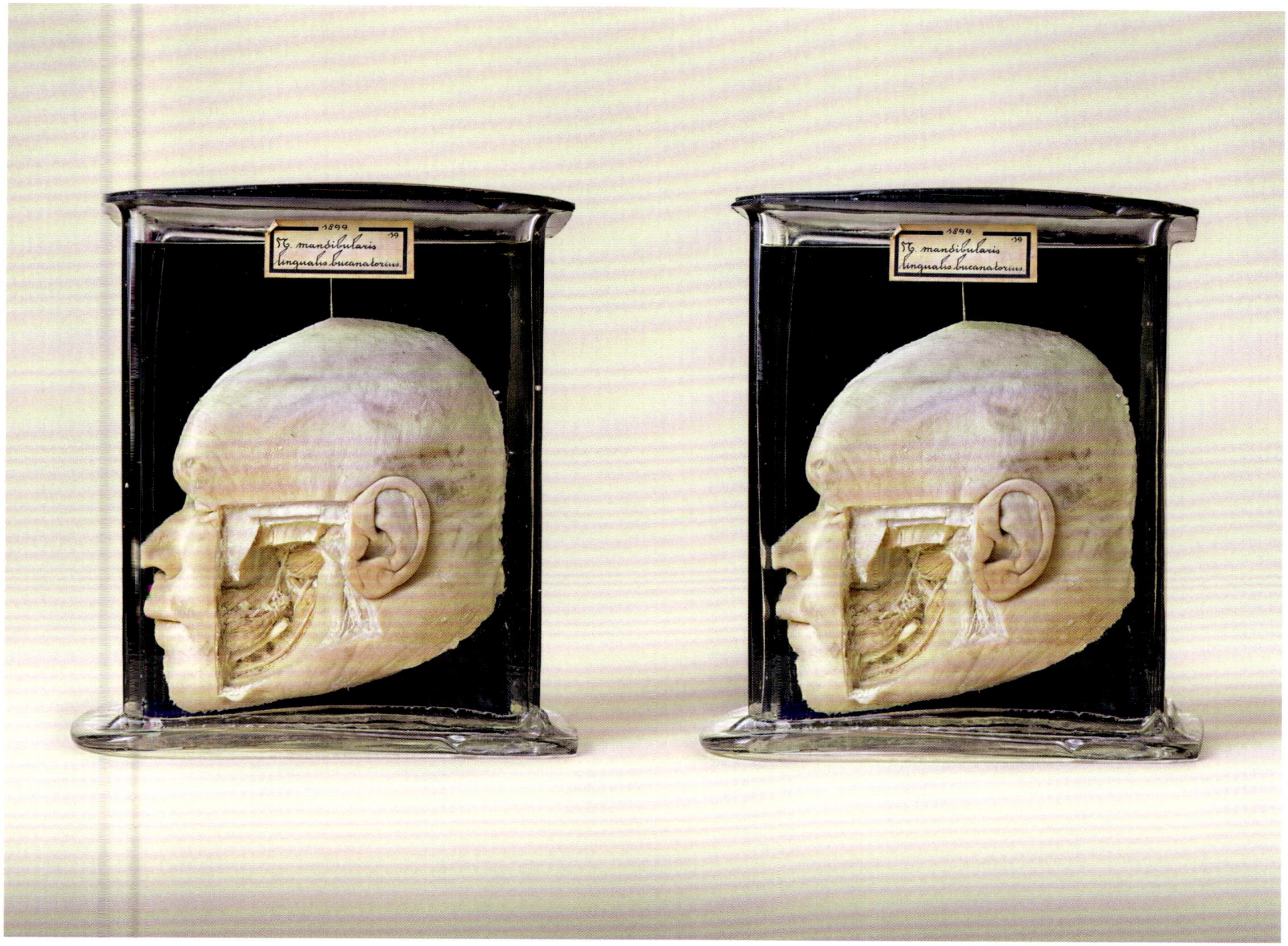
1899
39
N. mandibularis
lingualis bucanatorius
1899
39
N. mandibularis
lingualis bucanatorius

ADULT FACE SHOWING LACRIMAL DUCTS
1899

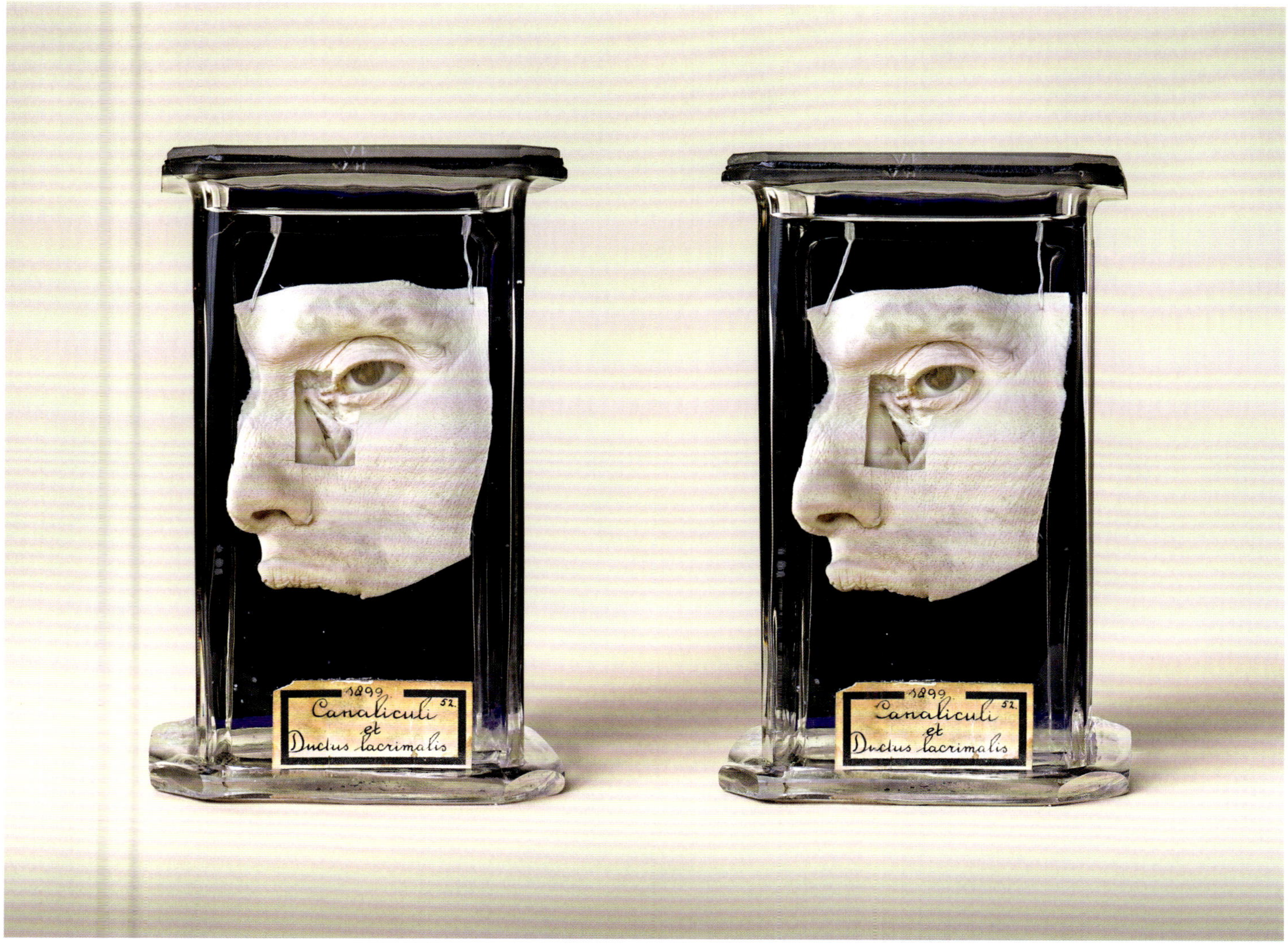
1899 52.
Canaliculi
et
Ductus lacrimalis
1899 52.
Canaliculi
et
Ductus lacrimalis

HEAD AND BRAIN OF A NEWBORN INFANT
1899

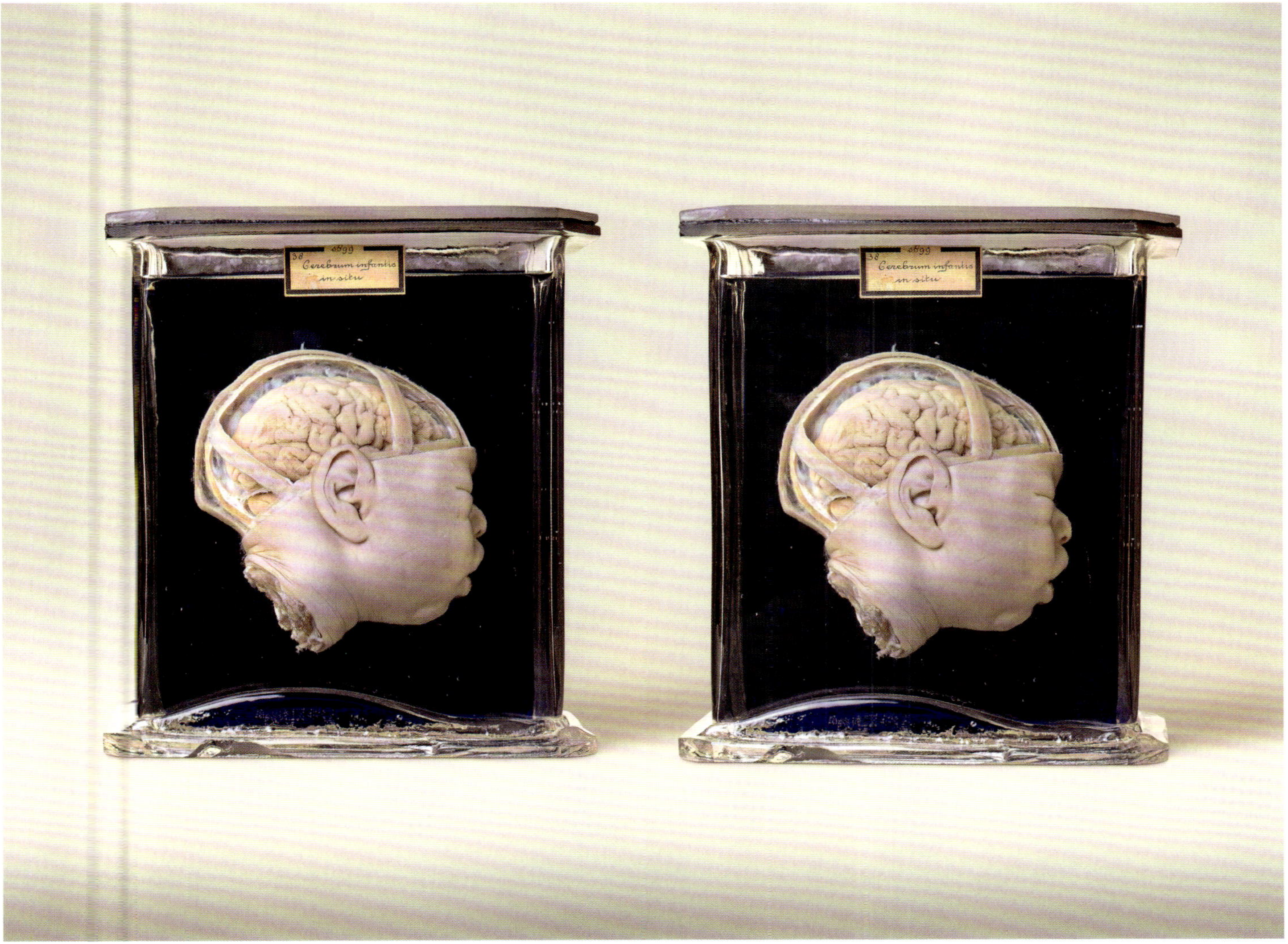
38 6899
Cerebrum infantis
in situ
38 6899
Cerebrum infantis
in situ

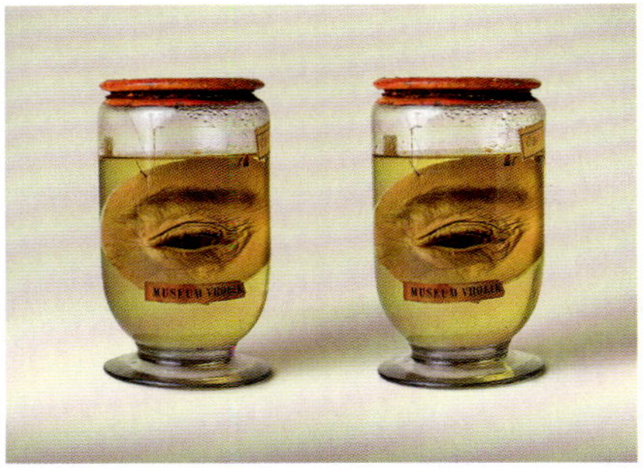

Eye and eyebrow
Before 1830
This anatomical preparation is special because it is one of very few specimens from the original Vrolik collection that remains fully intact. The lid, suspension method and liquid have not been changed since father or son Vrolik preserved the specimen some 180 years ago. This is extraordinary since both the suspension (a horse hair) and the lid (a round piece of slate stuck to the rim with sealing wax and covered by a pig's bladder) are very fragile. Even the preservation liquid – alcohol – normally evaporates over time.

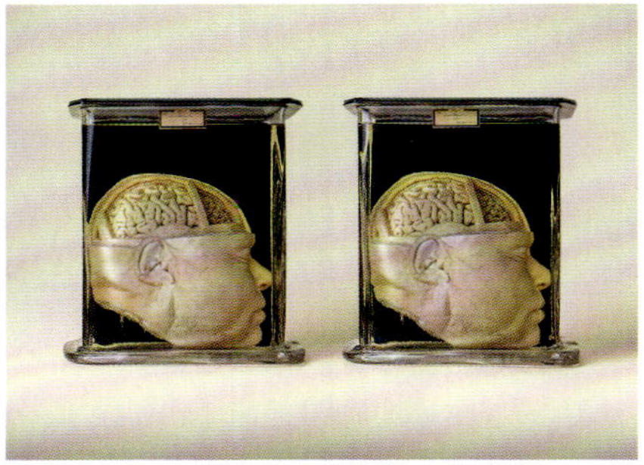

Head and brain of an adult male
1898
The cerebrum (the large brain) with its characteristic coils and grooves is clearly visible and accounts for around two-thirds of the weight of an average adult brain of about 1.5kg (3.3lb). The cerebrum forms the largest and uppermost portion of the brain and is where most complex processes take place.

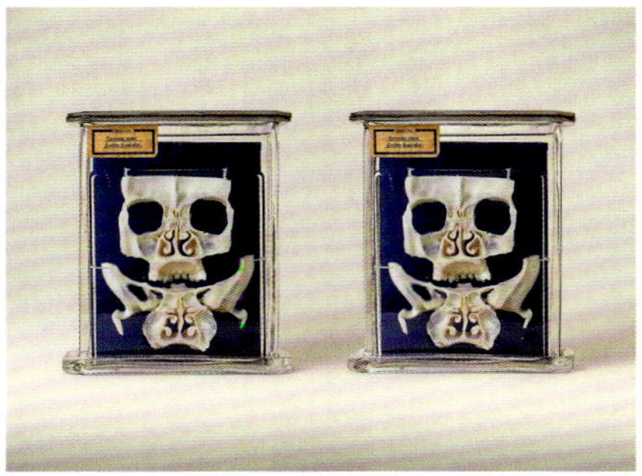

Two cross sections of the facial skeleton
1886
The top cross section shows the two orbits, below which are the two maxillary sinuses. In the middle of the face is the nasal cavity, with its upper chambers formed by nasal conchae (scroll-shaped, bony elements). The bottom cross section presents an area beyond the orbits and a little more to the back of the skull, but it is still possible to recognize the nasal conchae. In total, the nasal cavity contains three pairs of conchae covered with a humid mucous membrane and rich in blood vessels. Their job is to warm the air that is inhaled through the nose before it enters the lungs. The mucous membranes are inflamed when suffering with a cold.

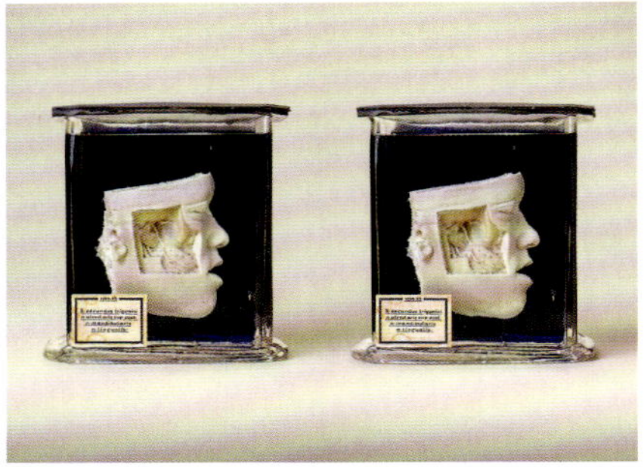

Head showing nerves of the cheek
1899
Part of the zygomatic bone and the mandible have been removed from the cheek area on the right side of this adult head. The preparation shows some of the sensory nerves of the oral mucosa, tongue, gums and teeth, and the motor nerves that allow movement in the jaw muscles. The thin, curving nerve in the middle is the superior alveolar nerve, the sensory nerve of the teeth of the maxilla. This nerve is usually anaesthetized by a dentist when extracting an upper molar or tooth.

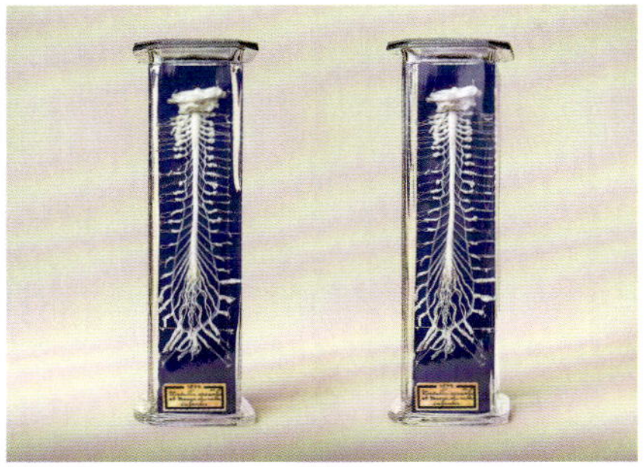

Spinal cord of an infant
1899
To the left and right of the spinal cord are the sympathetic trunks, two long chains of interconnected ganglia (clusters of nerve cells) that run along each side of the vertebral column from the base of the skull to the sacrum. The sympathetic trunks are fundamental parts of the sympathetic nervous system, which, in turn, is part of the autonomic nervous system that operates certain functions in the body without us having to think about them. For instance, when we start running it automatically increases the heart rate and the frequency of breath, and dilates the blood vessels in our muscles.

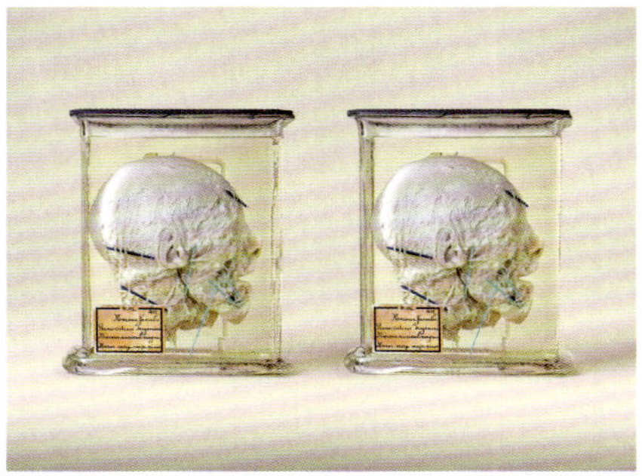

Superficial nerves of the head
Between 1863 and 1877
Dark- and light-blue glass rods are used to better differentiate the nerves from the rest of the whitish tissue on this complex network of superficial nerves of the head and face. Just below the earlobe is the facial nerve, the branches of which stimulate movement in most facial muscles. A paralysis in any of the branches of this nerve results in the loss of function in one or more of the facial muscles, causing, for example, the corner of the mouth to hang down.

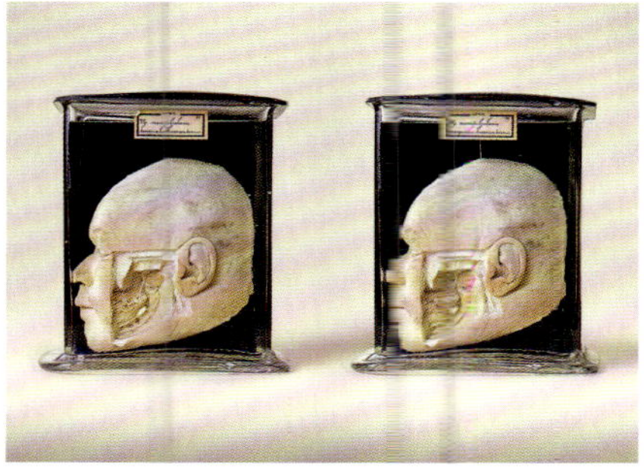

Left side of adult head showing nerves of the cheek
1899
Part of the zygomatic bone and the mandible have been removed to expose the mandibular nerve, among others. On the far right, it is possible to see the point at which this nerve enters the mandible. It is at this spot that a dentist, extracting a lower molar or tooth, would anaesthetize the nerve, causing all the teeth and gums in front to go numb.

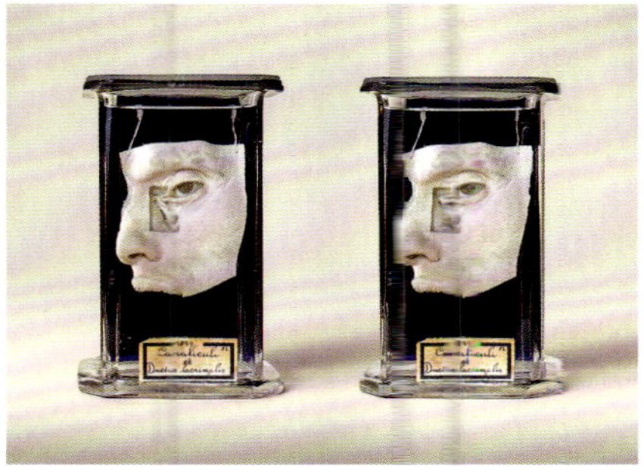

Adult face showing lacrimal ducts
1899
The skin of the inner corner of the eye has been removed to show the nasolacrimal ducts. Tears are generated in the outer, upper corner of the eye by the lacrimal gland. When crying, the eyelids press together in such a way that the tear fluid flows into the lacrimal sacs in the inner corner of the eye. From there, it enters the nasolacrimal duct that transports the fluid to the nose (this is why someone crying often starts to sniffle).

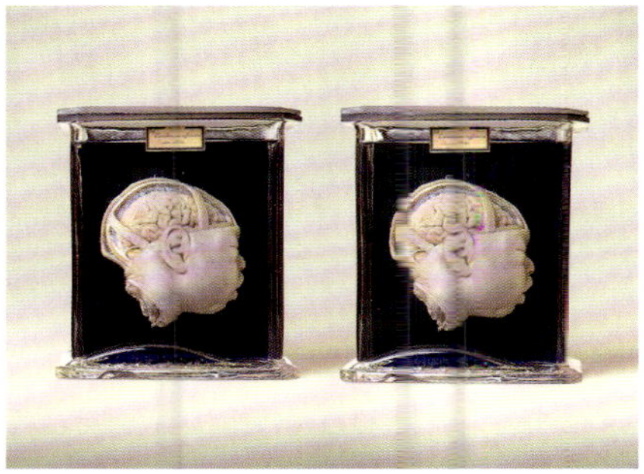

Head and brain of a newborn infant
1899
By making connections between neurons, the human brain continues to grow and develop until the age of 25 to 30, at which point it reaches full maturity. Around the age of ten, the brain starts to cut some connections and strengthen others to make itself more powerful and efficient.

Limbs

HAND WITH POLYDACTYLY
Before 1850

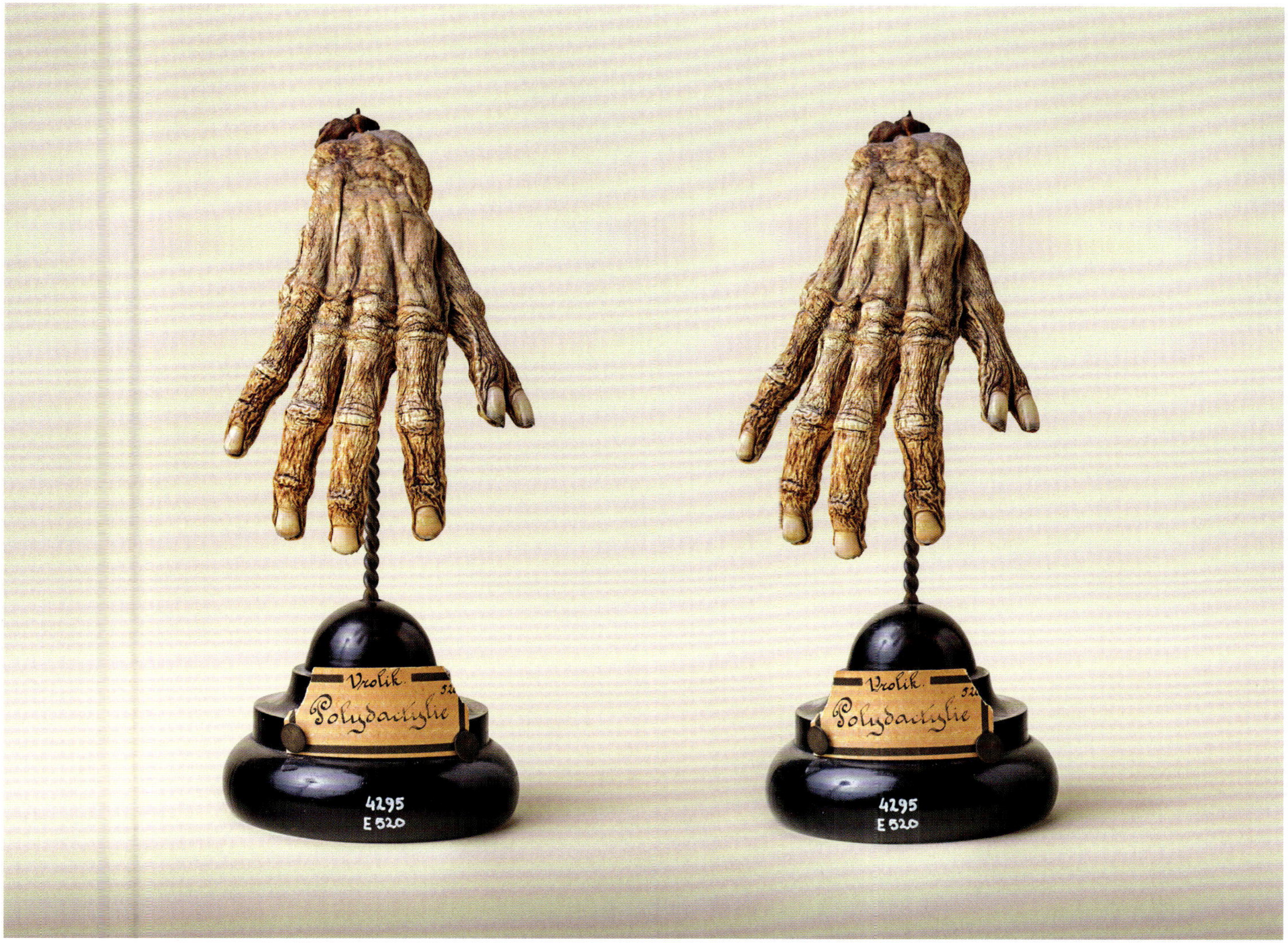
Vrolik.
Polydactylie
4295
E 520
Vrolik.
Polydactylie
4295
E 520

PLANTAR SIDE OF A CHILD'S RIGHT FOOT
1898

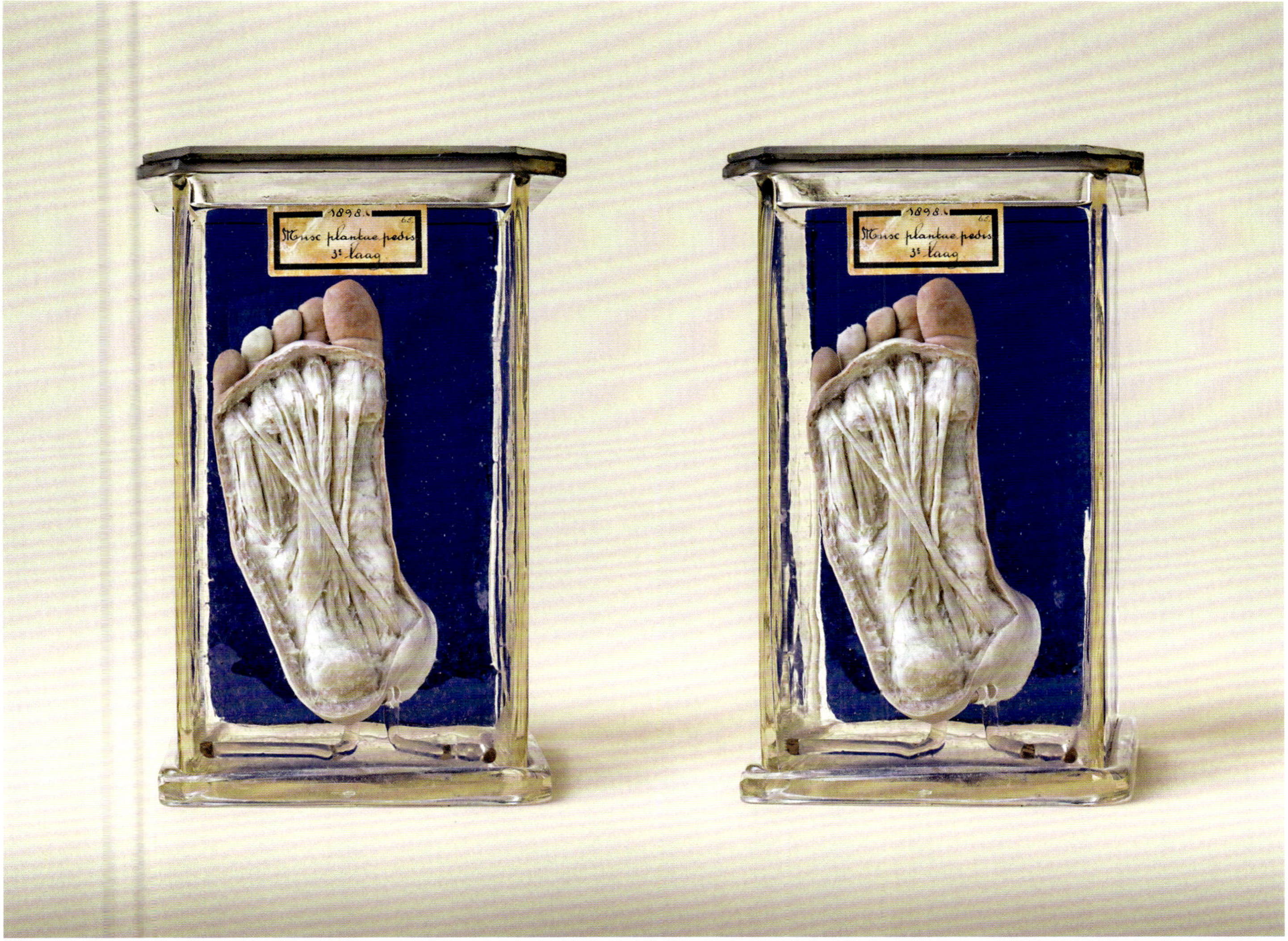
1898.
65.
Musc plantae pedis
3e laag

CROSS SECTION OF THE UPPER ARM
1899

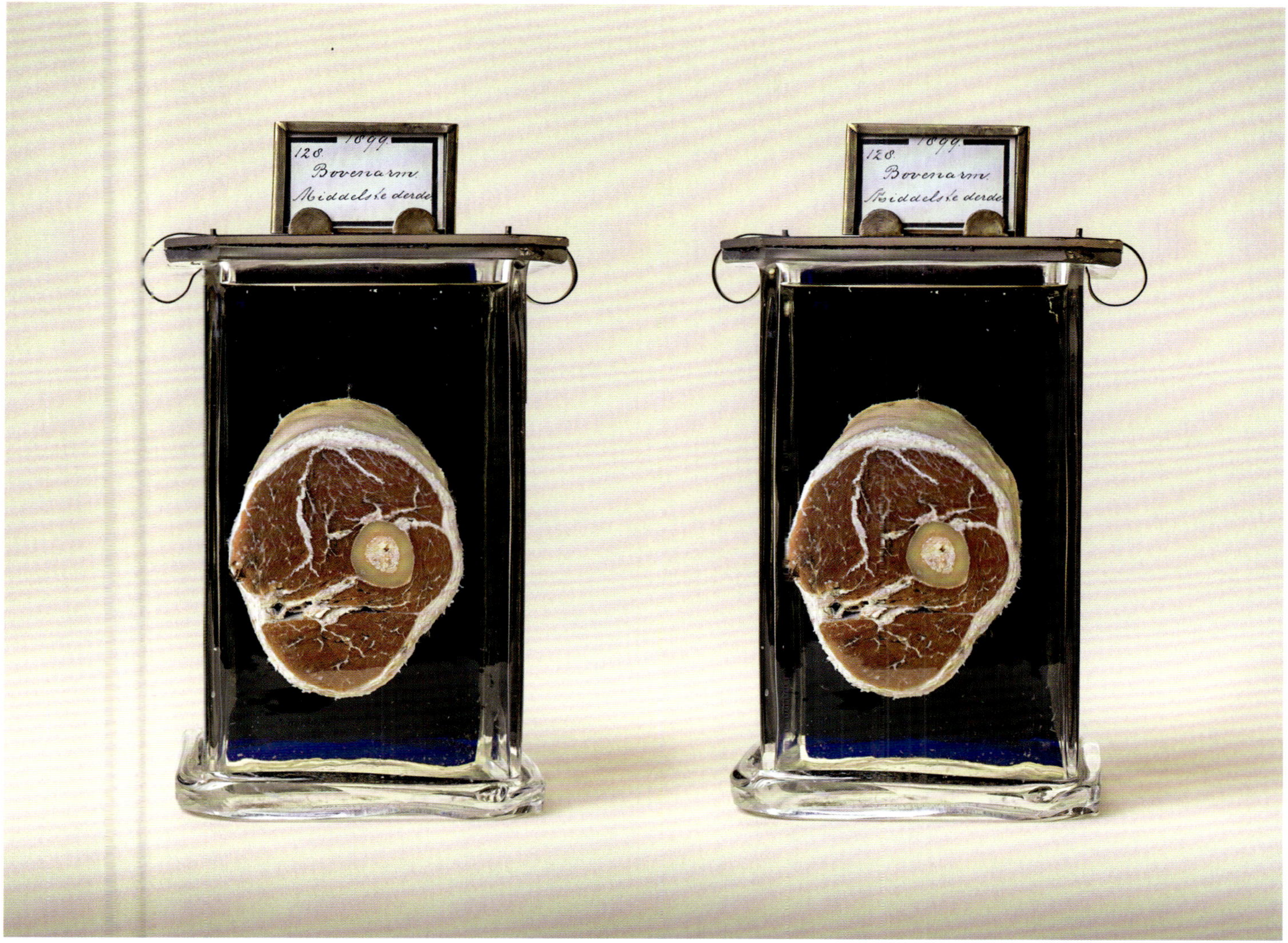
1899.
128.
Bovenarm
Middelste derde
1899.
128.
Bovenarm
Middelste derde

TATTOO TAKEN FROM THE UPPER ARM
1882

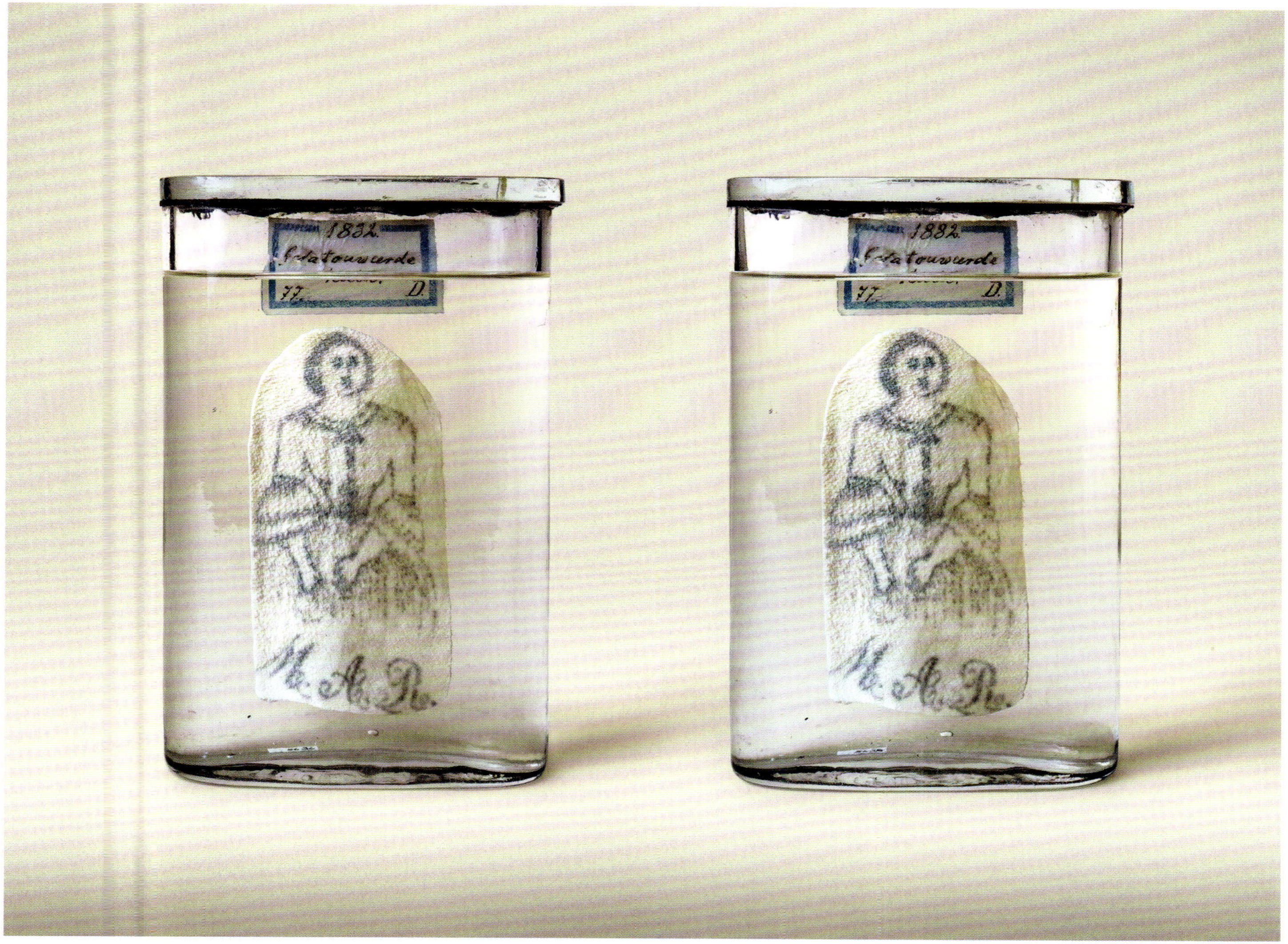
1882.
77.
D.
1882.
77.
D.

HUMERUS OF A SIX-YEAR-OLD BOY
1856–57

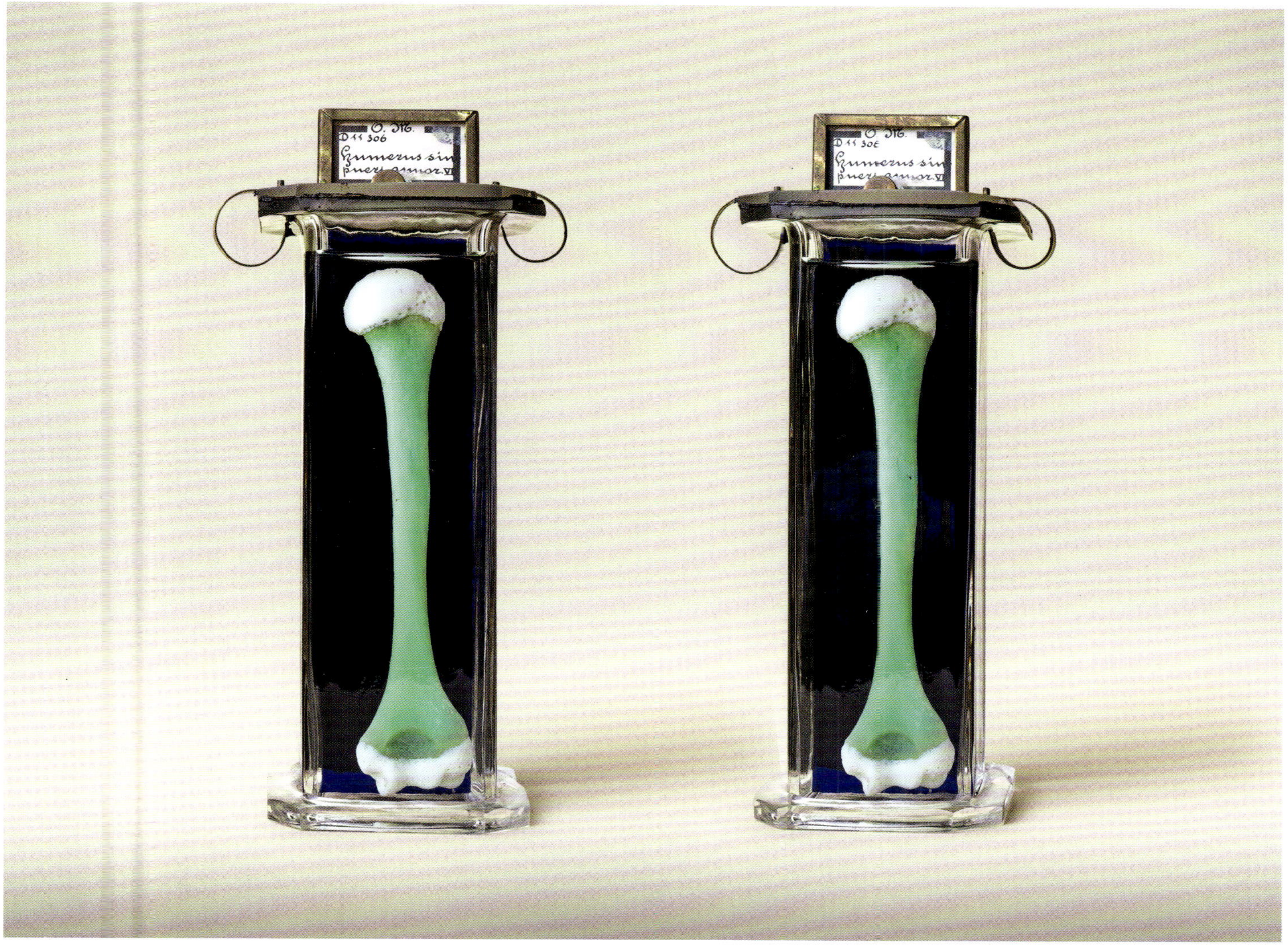
O. M.
D 11 306
Humerus sin
O. M.
D 11 306
Humerus sin

DORSAL SIDE OF A CHILD'S RIGHT FOOT
1899

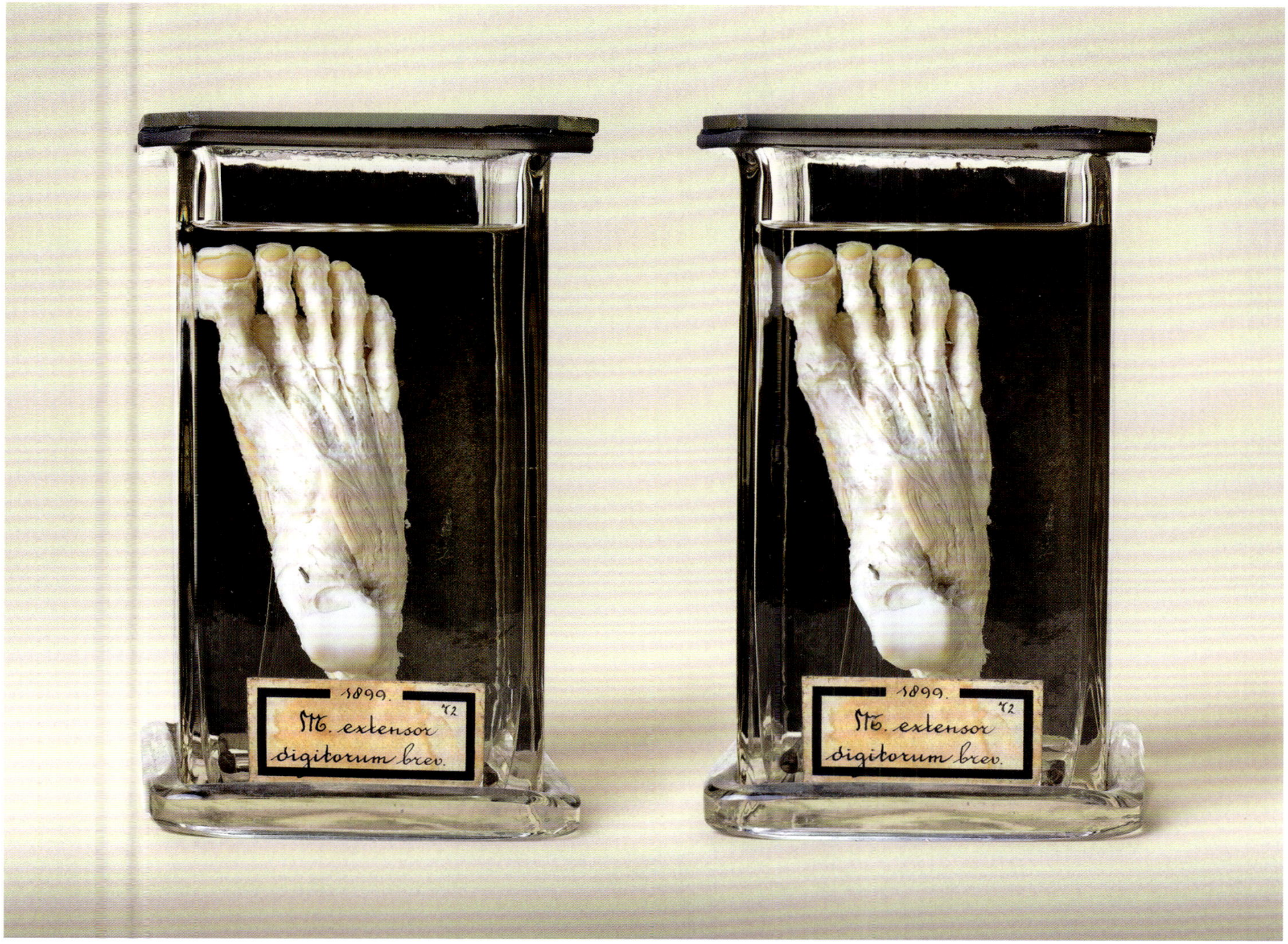
1899.
72
M. extensor
digitorum brev.
1899.
72
M. extensor
digitorum brev.

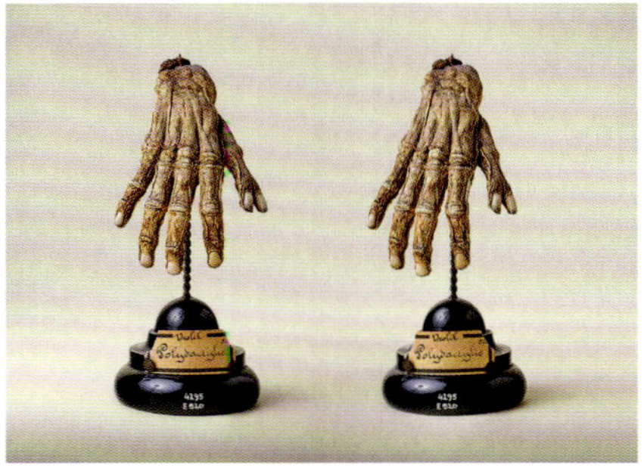

Hand with polydactyly
Before 1850
Polydactyly, or extra digits, in this case an additional thumb, can occur on hands and/or feet, and can be a symptom of a more complex syndrome or appear in isolation. In the latter case, it is sometimes hereditary, and the extra finger or toe nowadays is often surgically removed at a young age.

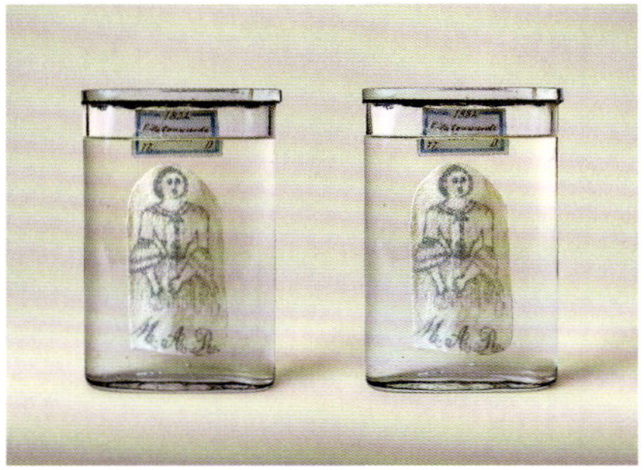

Tattoo taken from the upper arm
1882
This fragment of human skin from the upper arm shows the tattoo of a woman and her initials 'M.A.R.' Based on her dress, she must have lived during the nineteenth century, and it can only be assumed that she was deeply loved by the (most likely) man who had her immortalized on his skin. In the Western world at that time, it was usually sailors and soldiers who had tattoos. Examples of this sort have more historical than anatomical value.

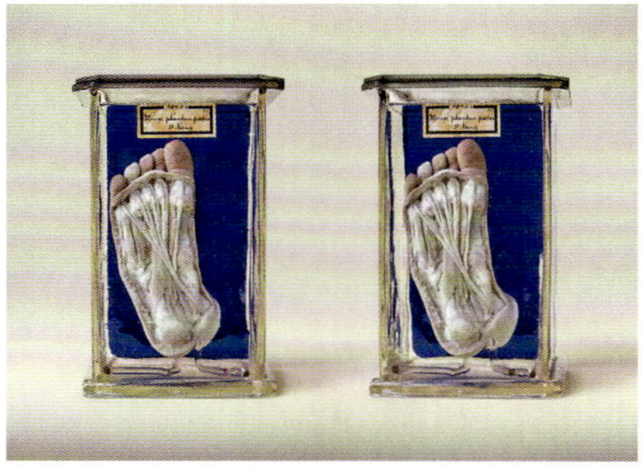

Plantar side of a child's right foot
1898
In the middle of the foot, the tendons of the long flexor muscle connect to the toes to allow the toes to bend. The long flexor muscle itself is not located on the sole of the foot but on the inner side of the calf, and can be felt working there when bending the toes.

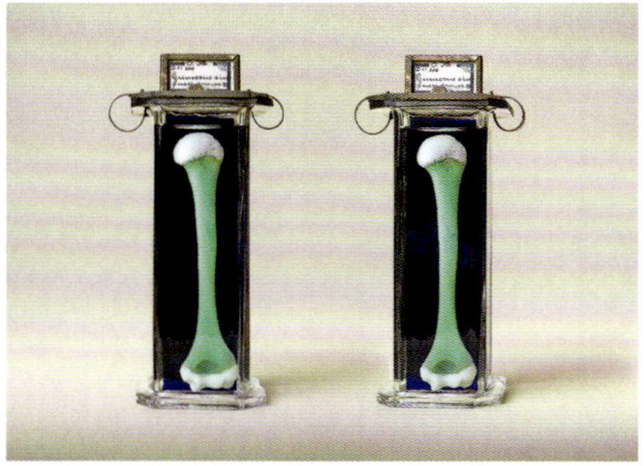

Humerus of a six-year-old boy
1856–57
The bony tissue is coloured with a green dye to distinguish it from the white cartilage. At the age of six, both ends of this bone are still largely composed of cartilage, with only a small ossification centre at their core. Between this ossification centre and the long ossified middle piece of the bone, a small layer of cartilage exists. When a bone grows, new bony tissue is added to the existing bone of the long middle part and the two ossification centres at the ends from these layers of cartilage.

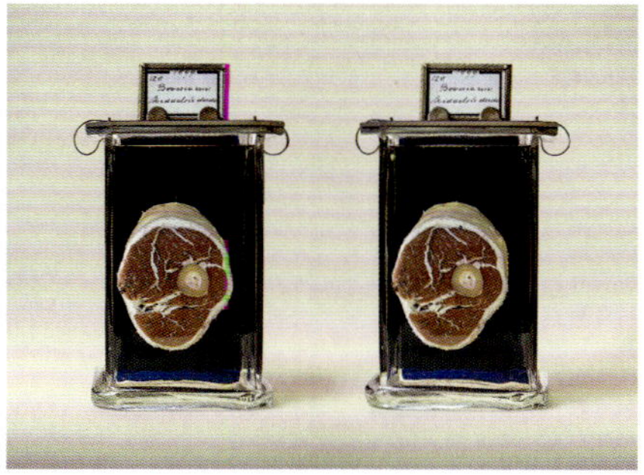

Cross section of the upper arm
1899
The muscles of this middle part of the upper arm are red-brown, and are surrounded by white layers of fat, tendons and fascia (sheaths of connective tissue that enclose all muscles and bundles of muscles). The bone on the middle right is the humerus, the marrow of which is white because of the fat.

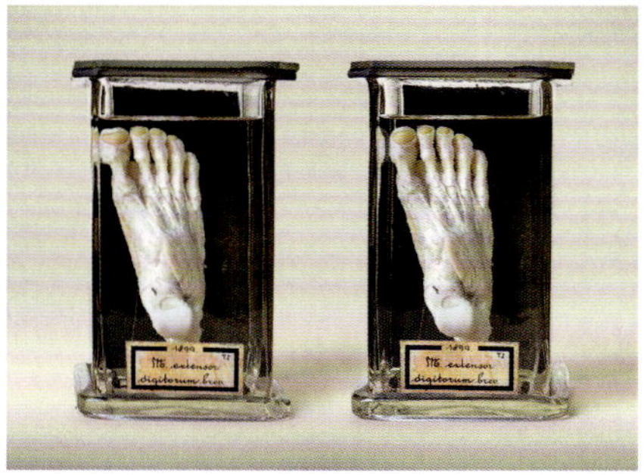

Dorsal side of a child's right foot
1899
The extensors – muscles that contract to extend or straighten a limb or body part – and their tendons are clearly visible on this child's right foot, and enable the toes to stretch. These muscles are the antagonists of the flexor muscles of the foot, which contract to bend the toes and are located on the plantar side of the foot.

Head & Neck

BLOOD VESSELS AND NERVES OF THE HEAD

1856–57

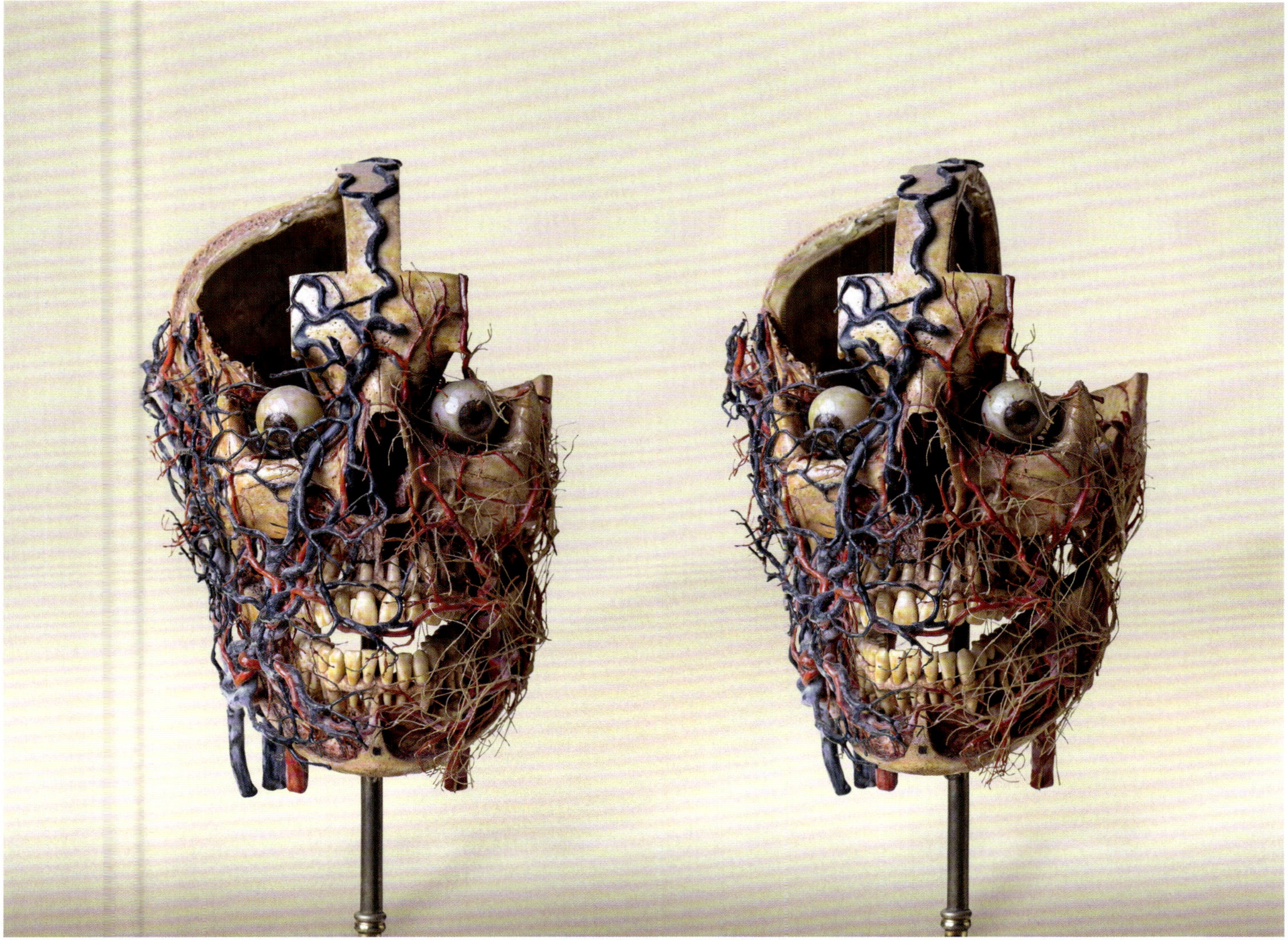

HEAD AND NECK SHOWING MUSCLES OF THE CHEEK, ORAL CAVITY AND PHARYNX
1893

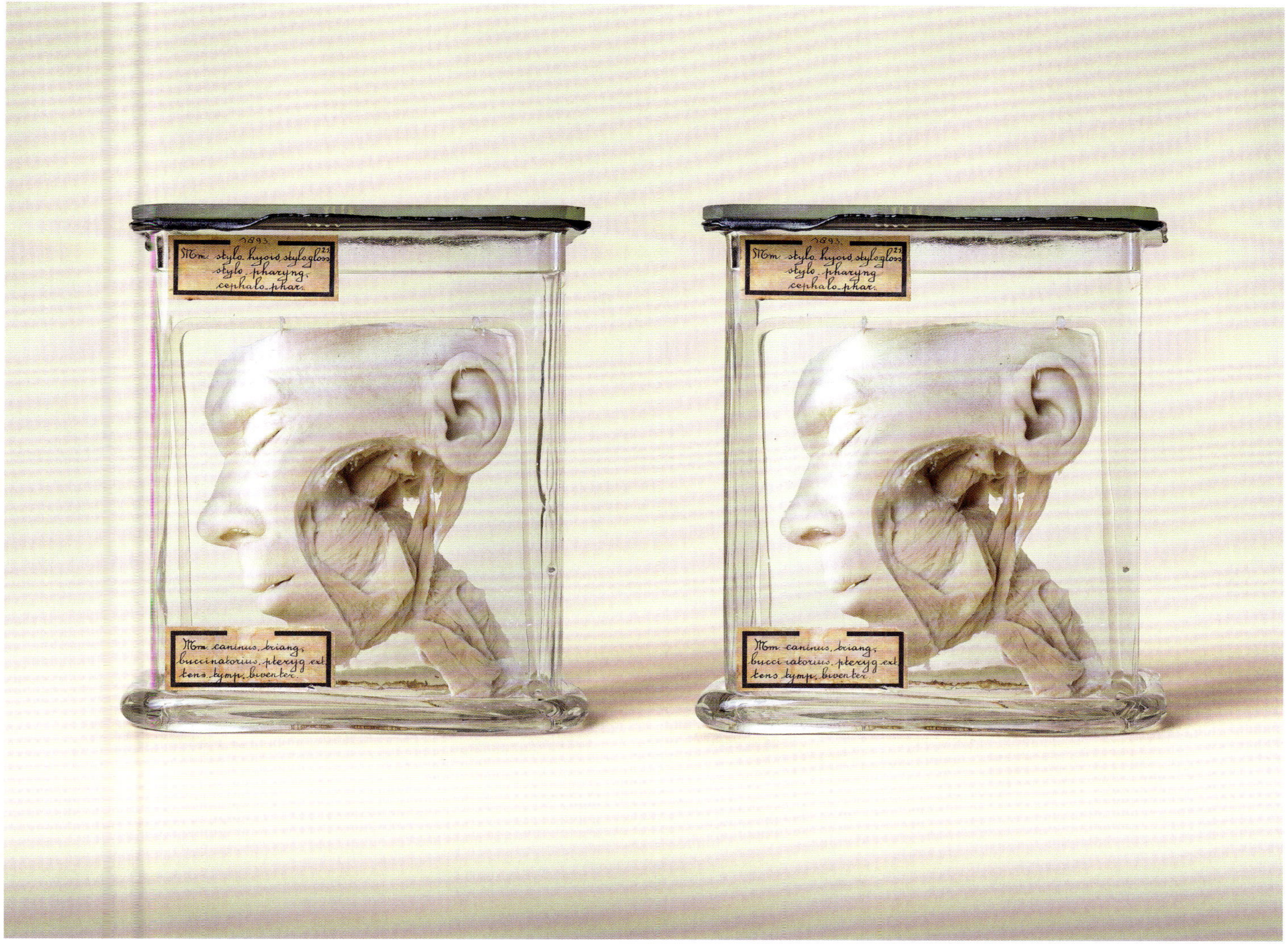

1893.
Mm. stylo. hyoid, stylogloss 25
stylo. pharyng.
cephalo. phar.
Mm. caninus, triang,
buccinatorius, pteryg. ext
tens. tymp. biventer.
1893.
Mm. stylo. hyoid, stylogloss 25
stylo. pharyng.
cephalo. phar.
Mm. caninus, triang,
buccinatorius, pteryg. ext
tens. tymp. biventer.

SKULL OF AN EIGHT-MONTH-OLD INFANT WITH HYDROCEPHALY
Before 1848

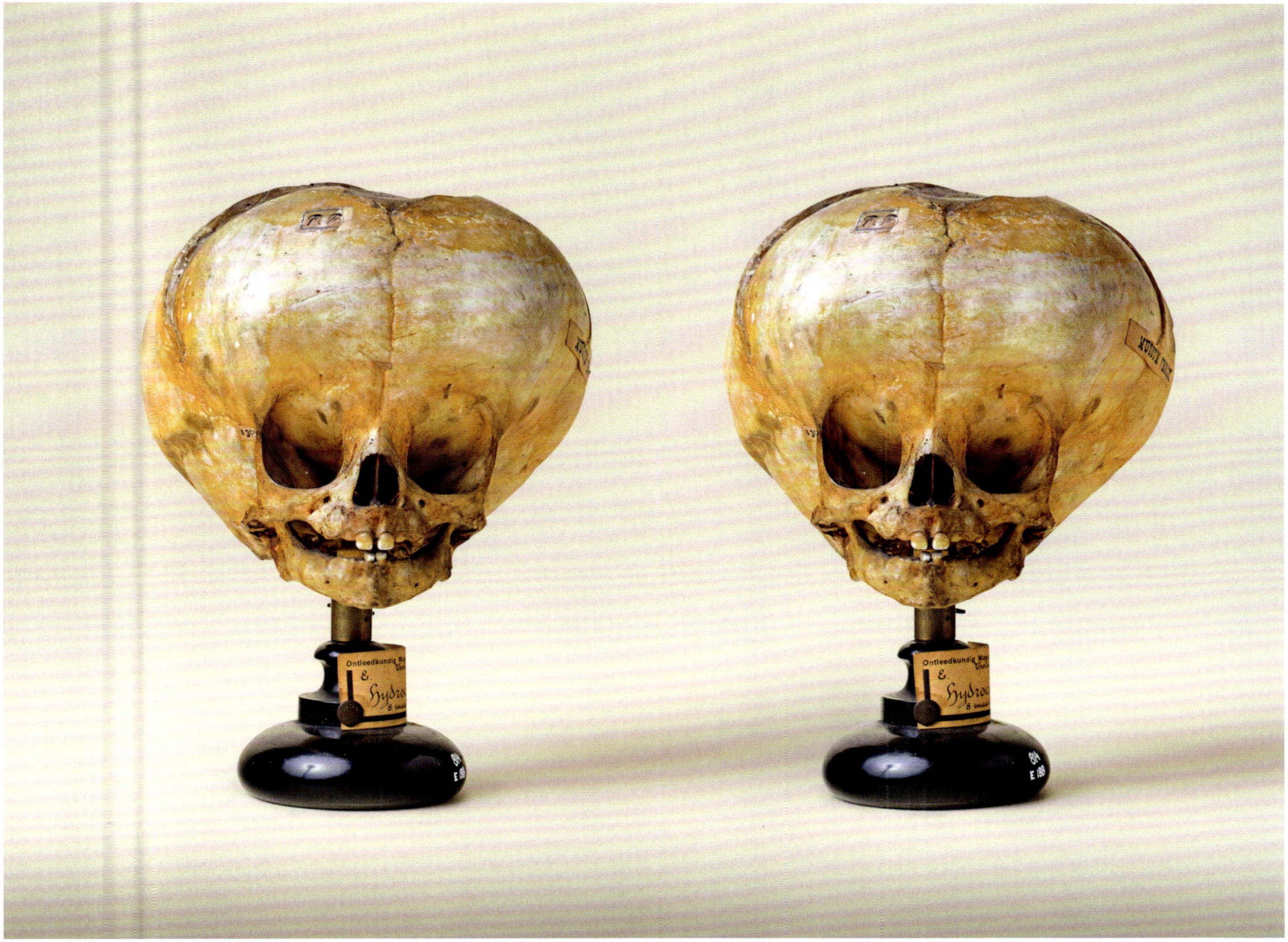

PART HEAD, PART SKULL OF AN ADULT MALE

1901

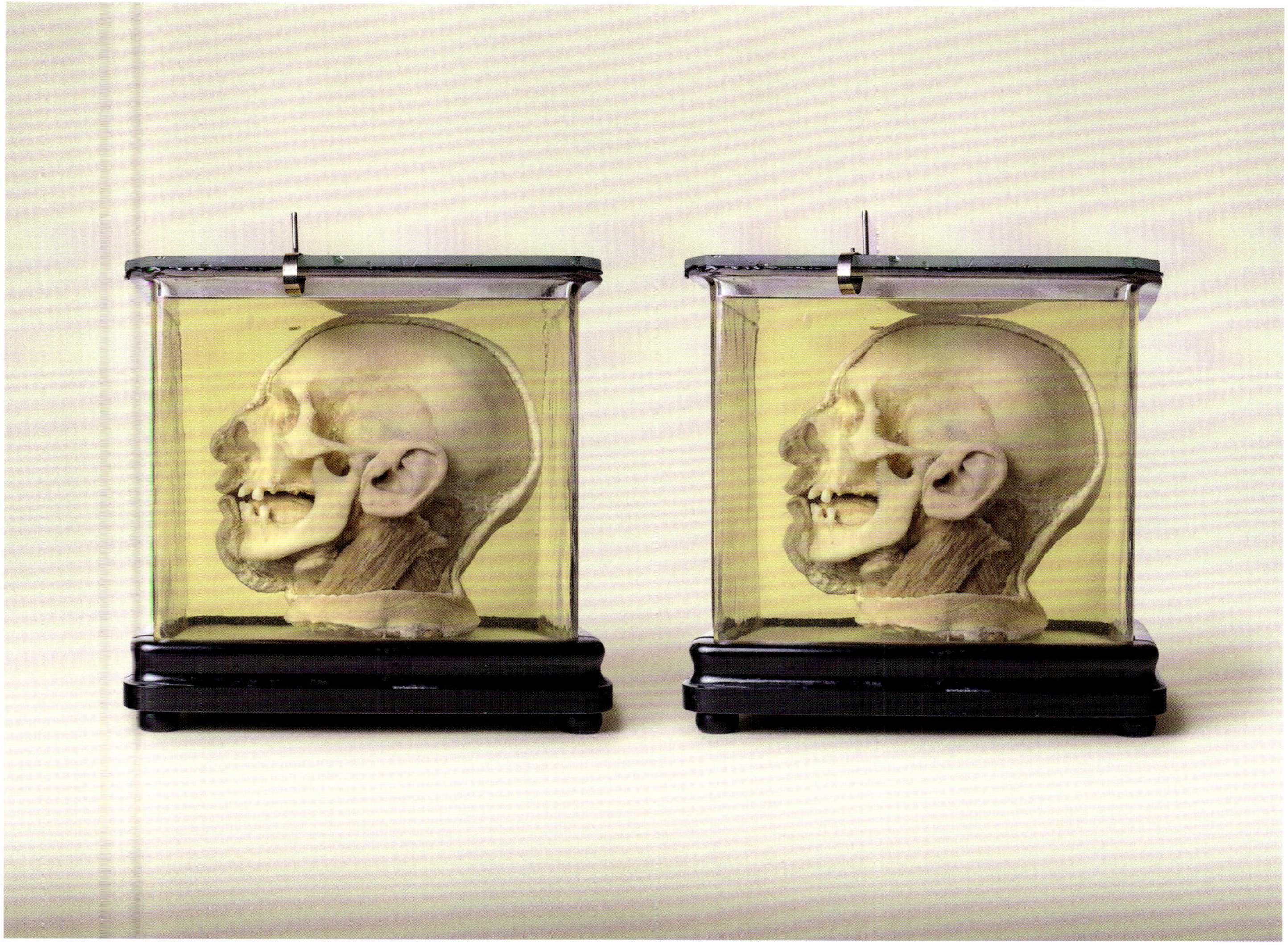

HEAD AND NECK SHOWING MUSCLES OF THE ORAL CAVITY AND PHARYNX

1881

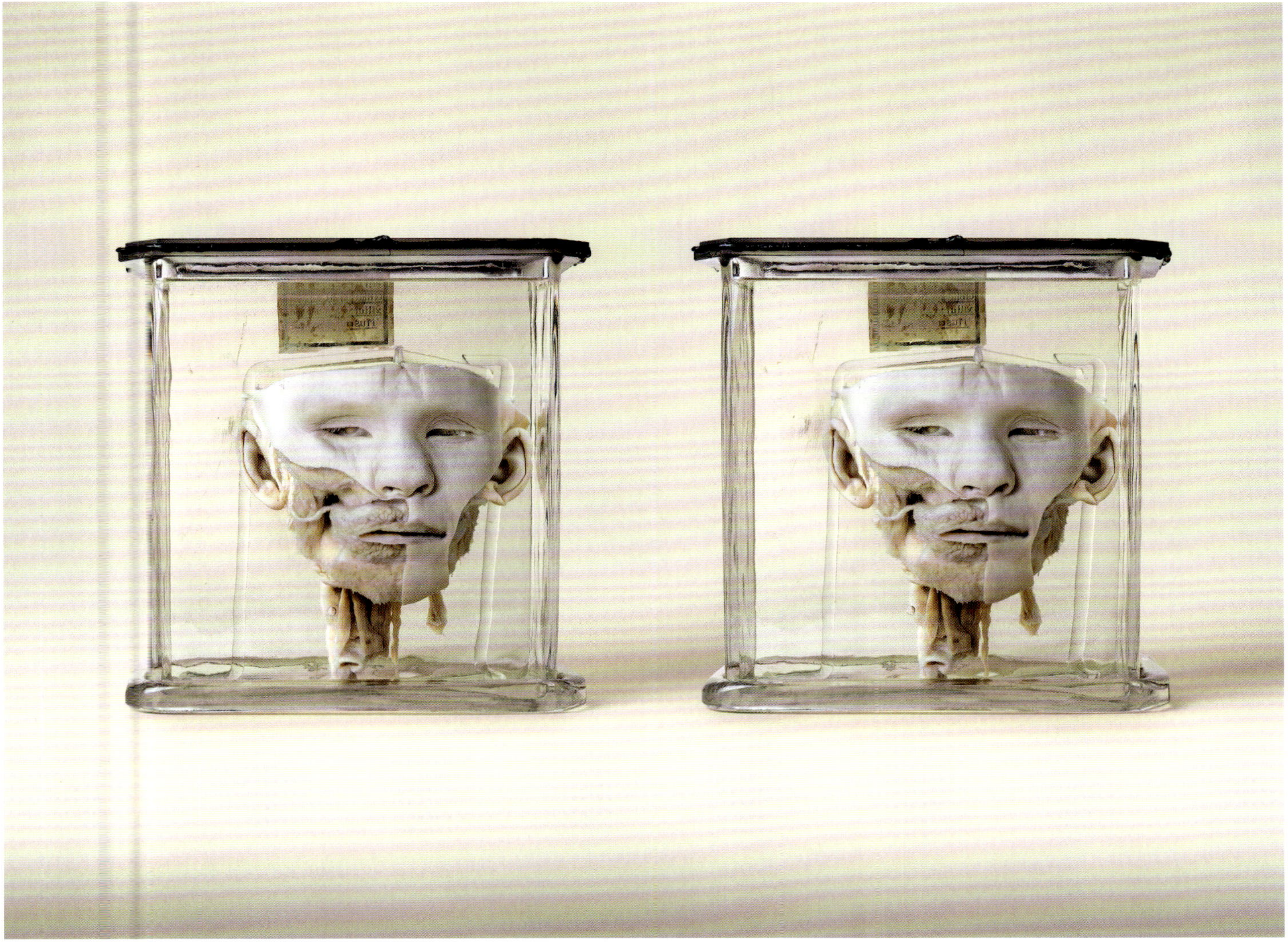

FACIAL MUSCLES OF THE HEAD

1892

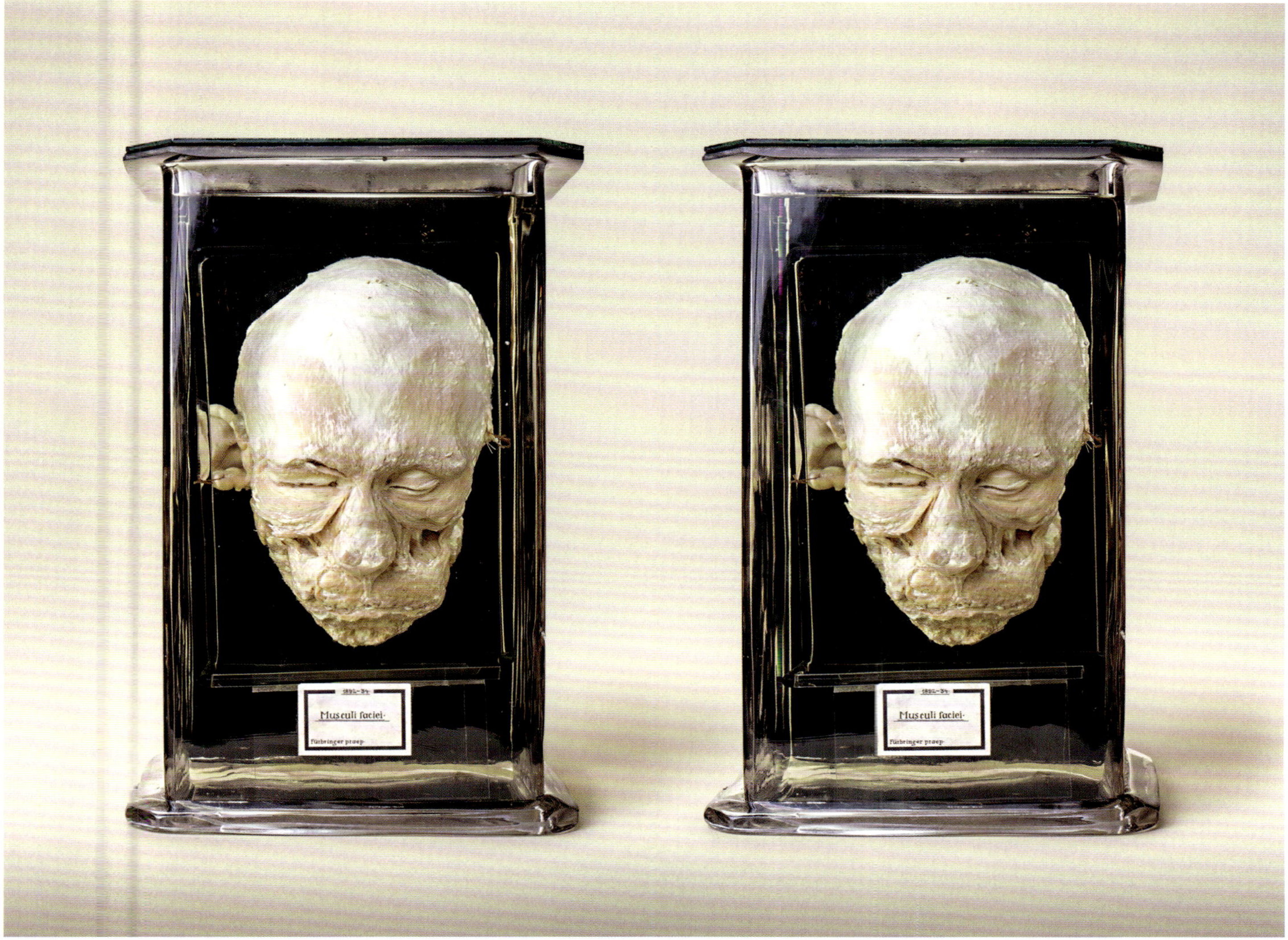
Musculi faciei·
Fürbringer praep·
Musculi faciei·
Fürbringer praep·

SKULL OF A CHILD SHOWING REMNANT OF THE METOPIC FONTANELLE
1914

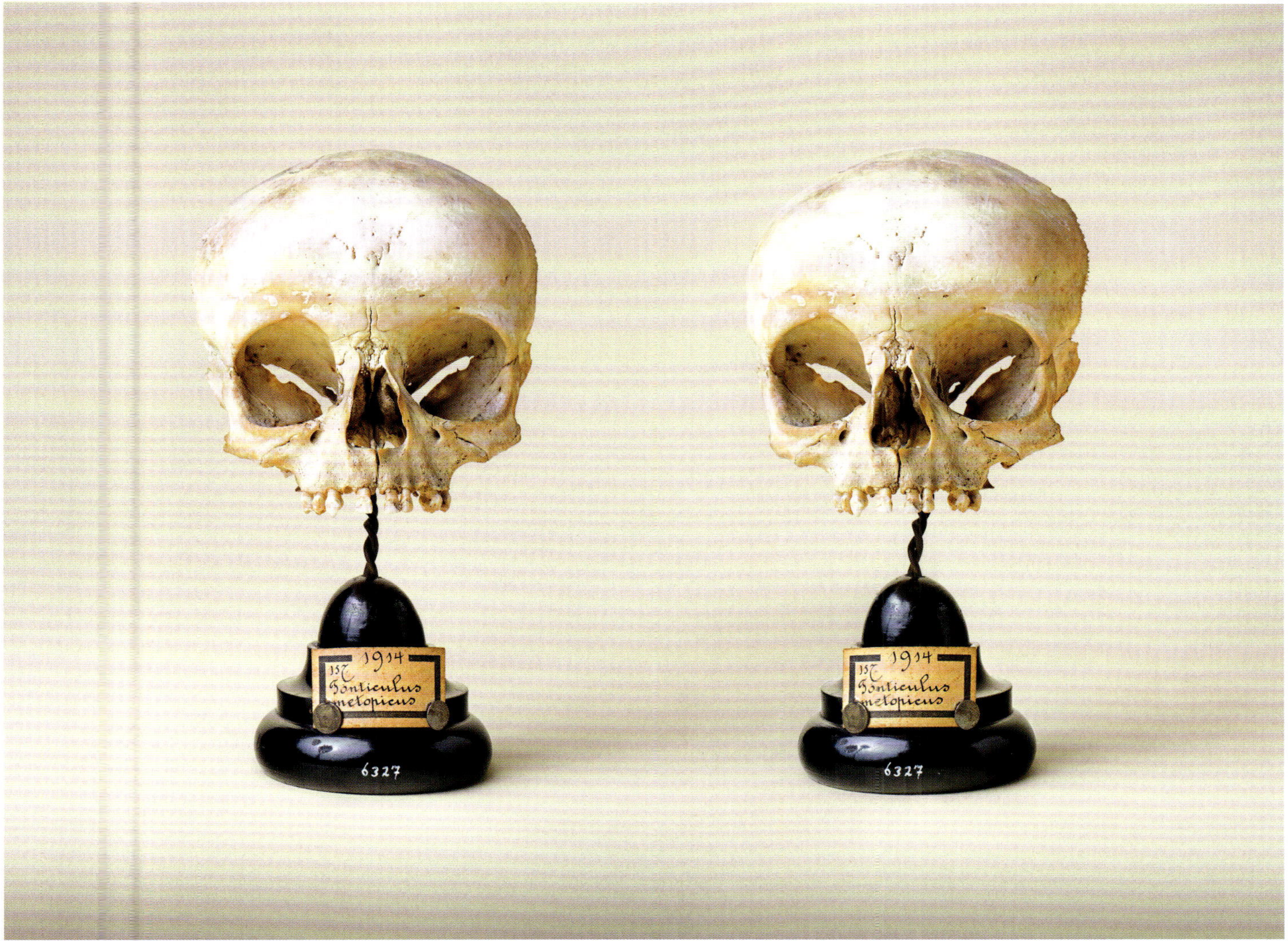
1914
Fonticulus
metopicus
6327
1914
Fonticulus
metopicus
6327

HEAD OF A CHILD SHOWING MUSCLES OF THE ORAL CAVITY

1892

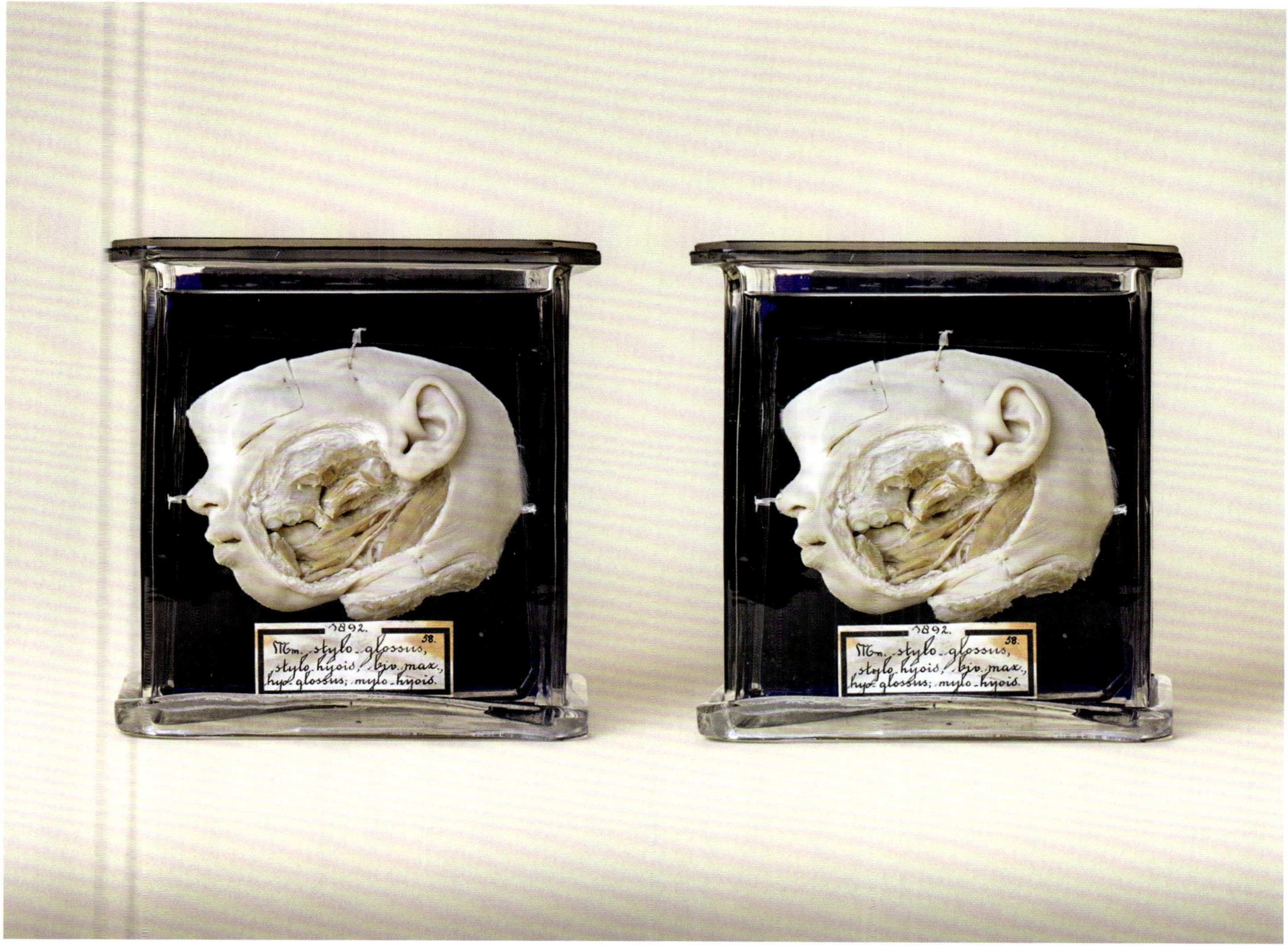
3892.
58.
Mm. stylo-glossus,
stylo-hyois, biv. max.,
hyo-glossus; mylo-hyois.

SUPERFICIAL ARTERIES OF THE HEAD
Second half of the nineteenth century

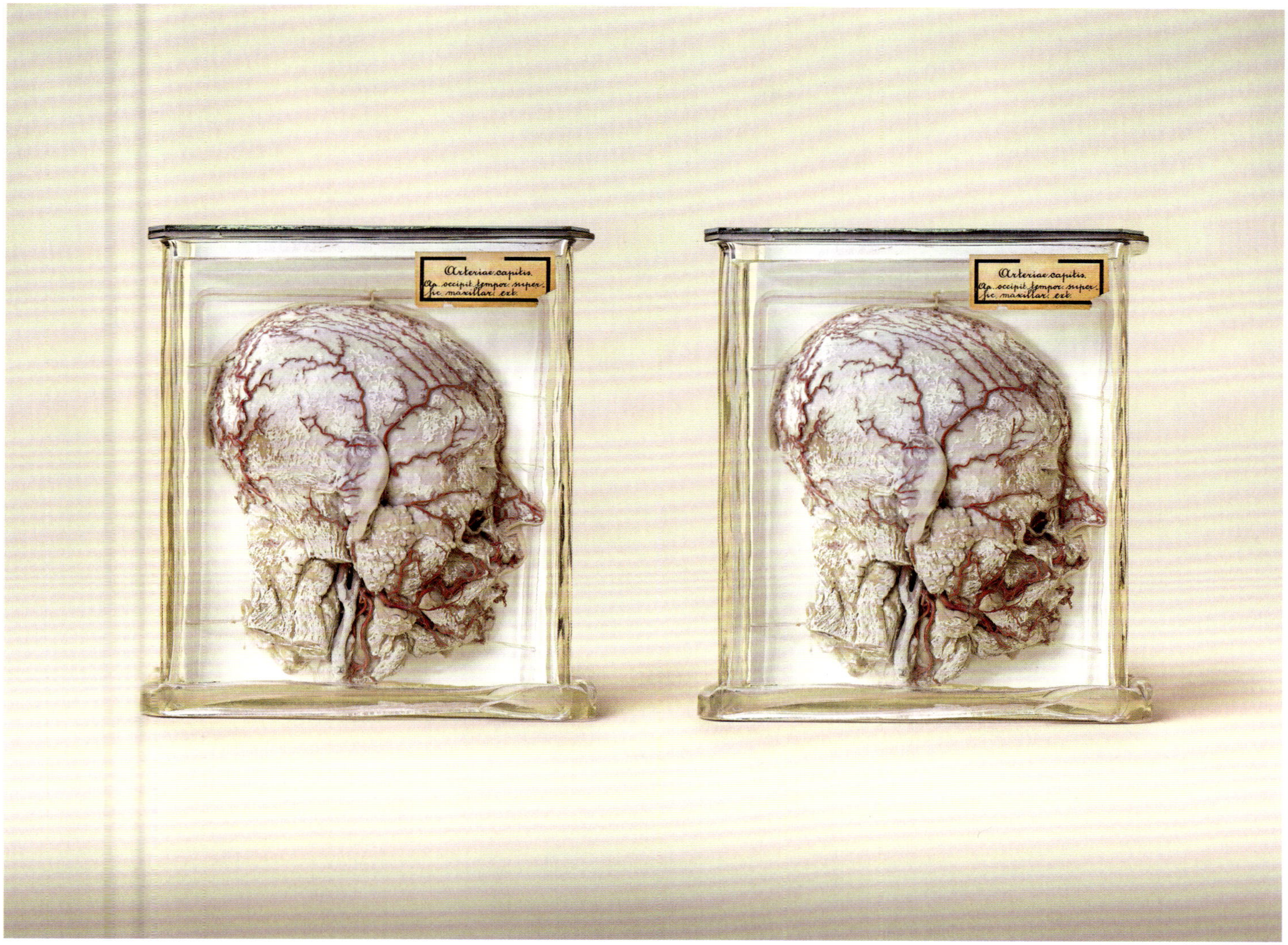
Arteriae capitis.
Aa. occipit. tempor. super-
fic. maxillar. ext.
Arteriae capitis.
Aa. occipit. tempor. super-
fic. maxillar. ext.

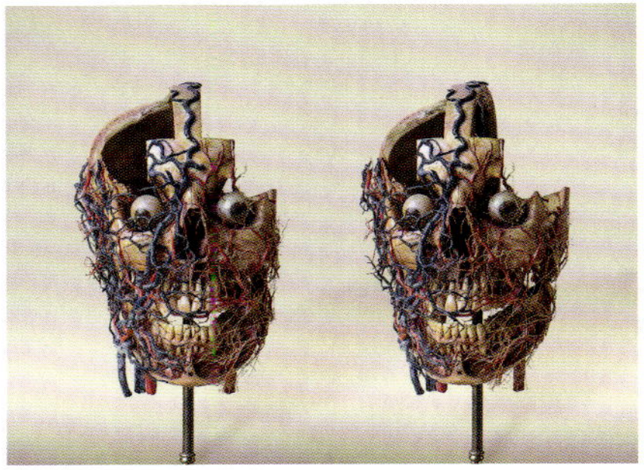

Blood vessels and nerves of the head

1856–57

The skull has attached to it artificial veins (blue wax), arteries (red wax) and nerves (yellow wax). Wax eyes have also been fitted. It was bought from the workshop of Vasseur in Paris, who created many anatomical models and specimens for institutions around the world. Willem Vrolik bought this example for educational purposes in the anatomical theatre of Amsterdam in 1856 or 1857.

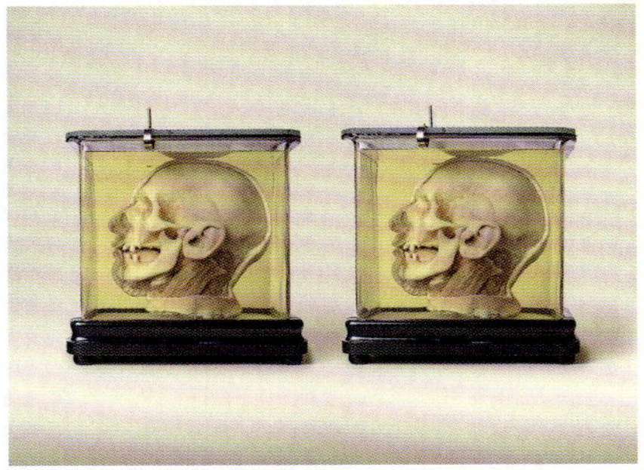

Part head, part skull of an adult male

1901

This lateral aspect of the head shows how the layer of soft tissues – muscle, fat, skin – fits over the skull. On the crown of the head, this layer is less than a centimetre thick. The ear, the tongue and the muscles of the neck have not been removed from the left side.

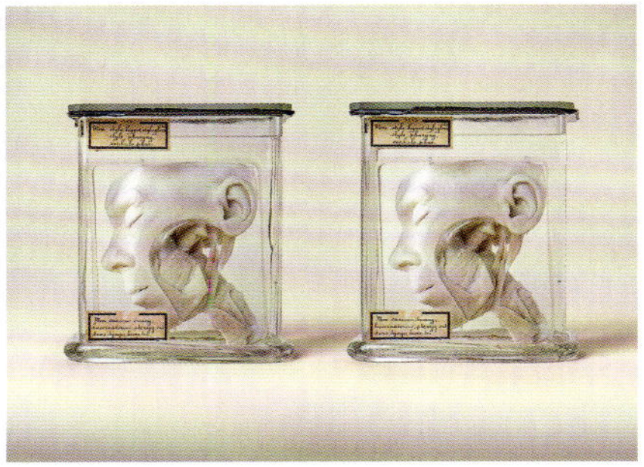

Head and neck showing muscles of the cheek, oral cavity and pharynx

1893

This fragment of the left side of the head has part of the mandible removed to show some of the muscles involved in swallowing and chewing. In the middle of the image, above the remaining part of the mandible, is the buccinator, or trumpeter, muscle. It is used by infants to suckle, and, later in life as the name suggests, for trumpet playing.

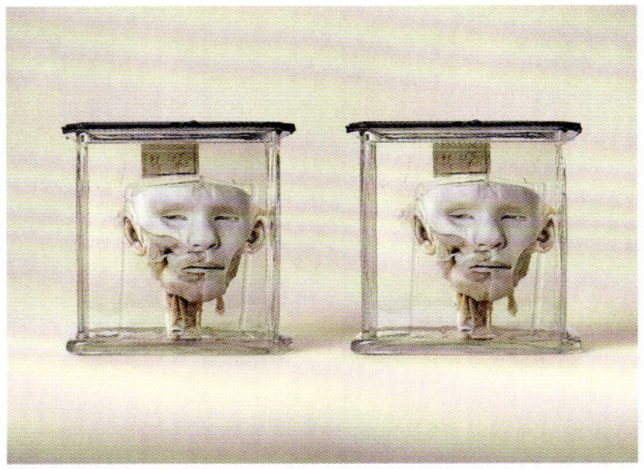

Head and neck showing muscles of the oral cavity and pharynx

1881

The skin of the cheek, chin, lip and neck has been partially removed to show the different muscles of the oral cavity and pharynx. The main purpose of these muscles is to swallow and chew. Below the ear is the parotid (meaning 'near the ear') gland, the major salivary gland of the mouth. It secretes saliva, which enters the oral cavity through the parotid duct. The duct is clearly visible on the left of this preparation, having the appearance of a white thread.

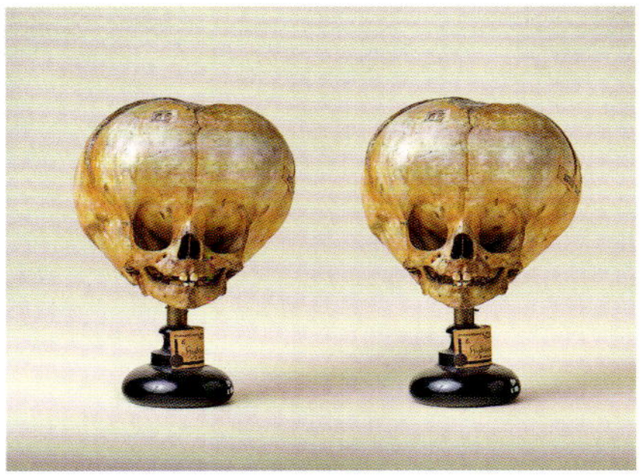

Skull of an eight-month-old infant with hydrocephaly

Before 1848

Hydrocephaly occurs when the cerebrospinal fluid produced in the brain cannot flow to the spinal cord. It is often caused by the blockage of the cerebral aqueduct, a narrow duct between the ventricles of the brain and the central canal of the spinal cord. Gradually the ventricles of the brain accumulate fluid, pushing the brain outwards and causing the skull to expand. In this case, expansion happened quite fast: not only have the cranial bones broadened, but the large fontanelle on top of the skull has also grown significantly (by around the age of eight months, this fontanelle should have diminished in size).

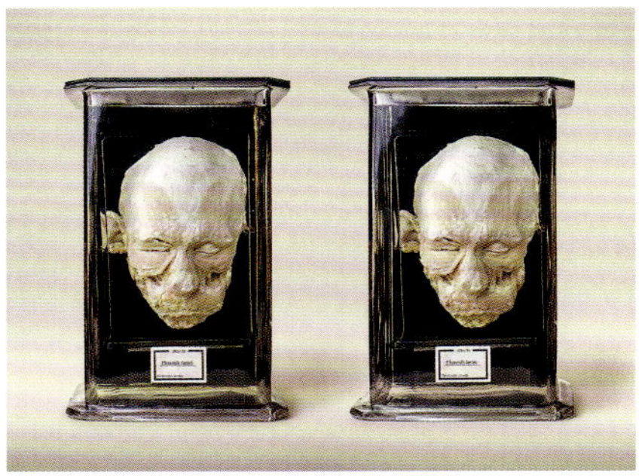

Facial muscles of the head

1892

The skin has been removed to reveal the superficial muscles of the face, which are also called mimic muscles because they are responsible for our facial expressions. The muscles surrounding the eyes are clearly visible, and, for example, allow us to squint in bright sunlight.

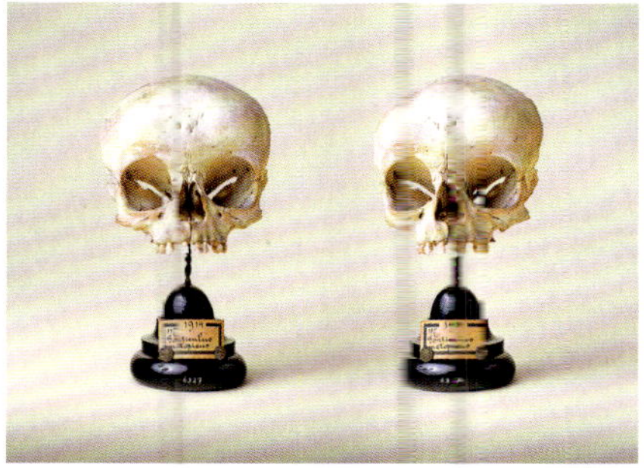

Skull of a child showing remnant of the metopic fontanelle

1914

The metopic fontanelle is a small fontanelle (gap) that in rare cases exists in the frontal suture, the fibrous joint between the two halves of the frontal bone in foetuses and newborns. This suture normally closes in the first year of life, creating the frontal bone. A metopic fontanelle is rare, and it is unusual to see a remnant of it in the skull of an older child.

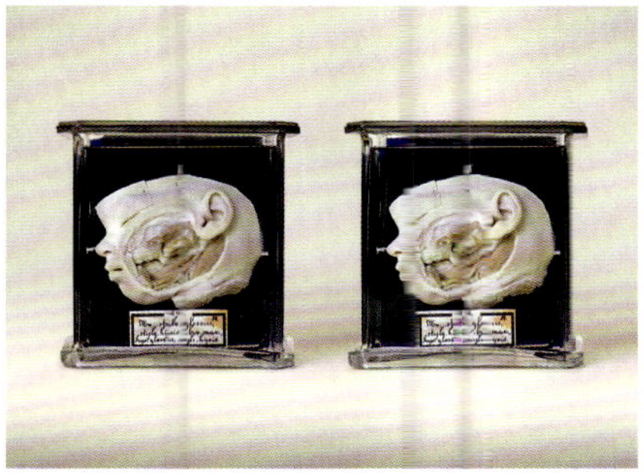

Head of a child showing muscles of the oral cavity

1892

The different muscles of the jaw, tongue and the floor of the oral cavity are revealed in this fragment of the left side of a child's head. The mandible and zygomatic bone have been partially removed. Several thin muscles – the digastric muscle, the stylohyoid muscle and the styloglossus muscle – run from underneath the ear to the lower jaw, where they attach to the tongue, the hyoid bone and the mandible. These muscles are all used primarily for the movement of the tongue and swallowing.

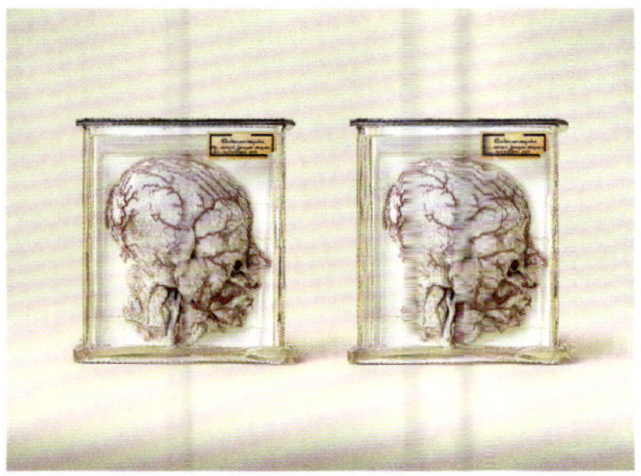

Superficial arteries of the head

Second half of the nineteenth century

The superficial arteries of the head, external to the cranium, have been injected with a red wax-like substance. These arteries originate from the pair of external carotid arteries, the right one of which can be seen in the neck of this preparation.

Internal Organs

ADULT HEART SHOWING THE CORONARY ARTERIES
Twentieth century

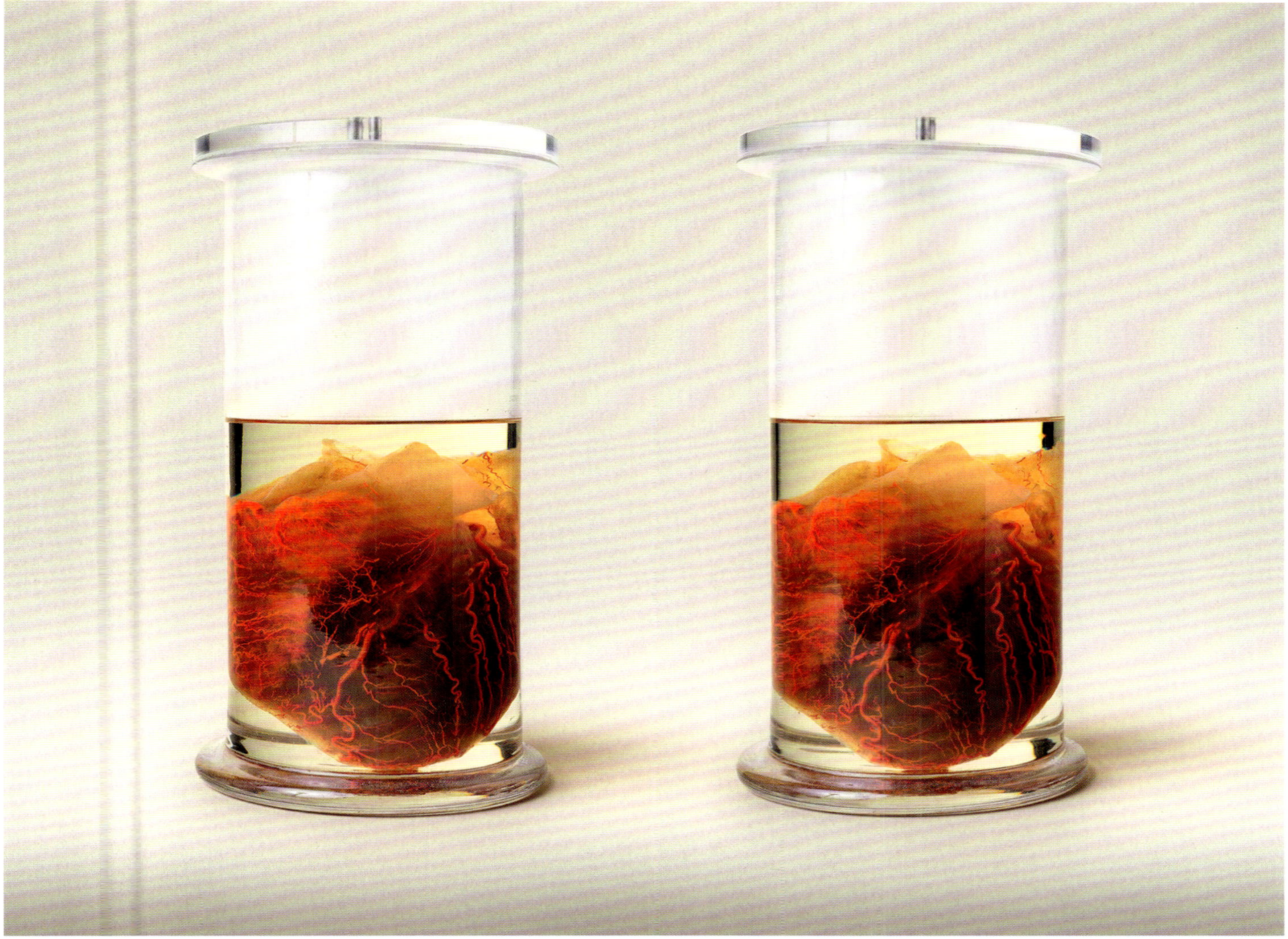

HORSESHOE KIDNEY
1884

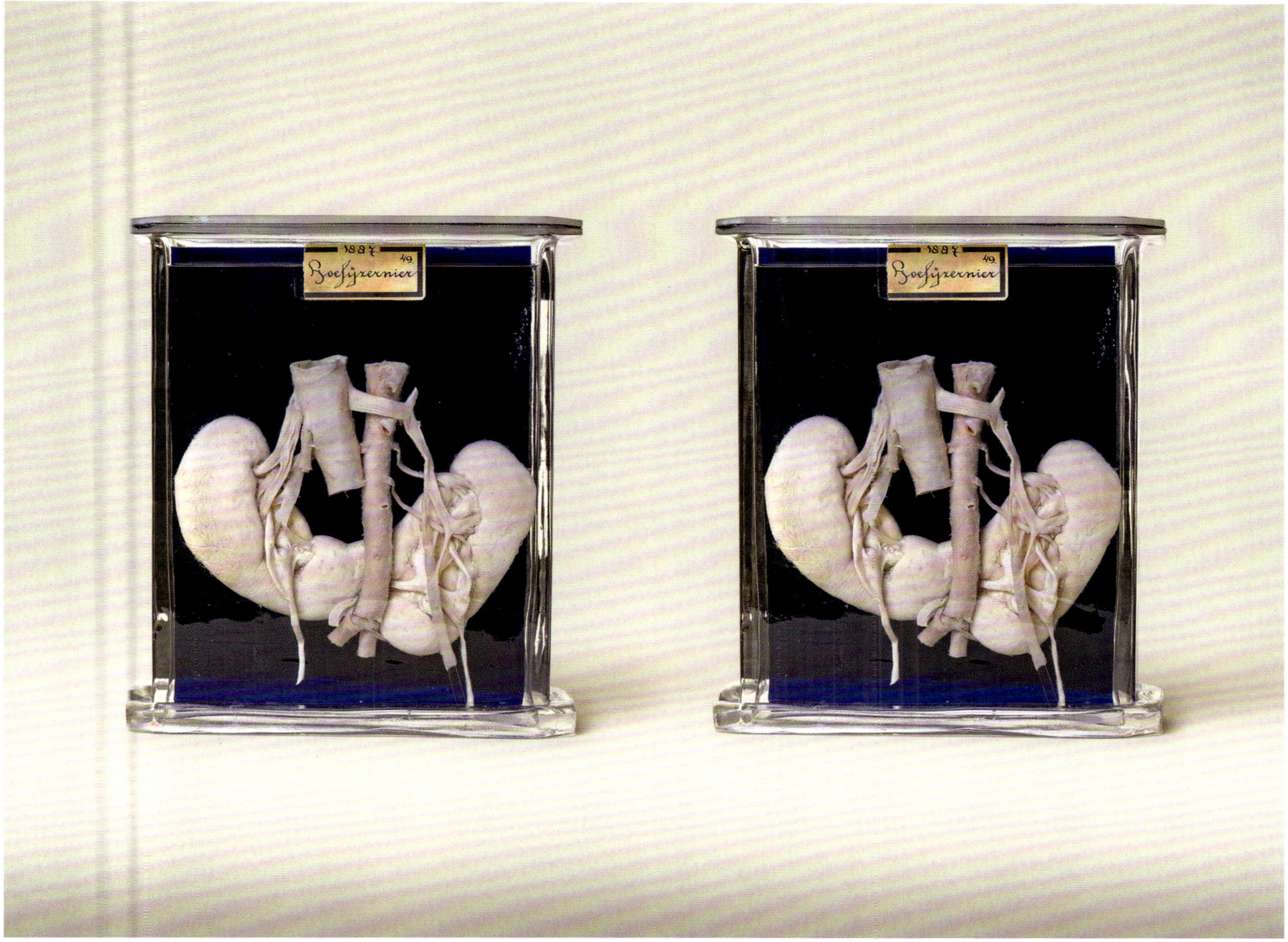
1887
49.
Hoefijzernier

HEART OF A YOUNG ADULT MALE
Before 1830

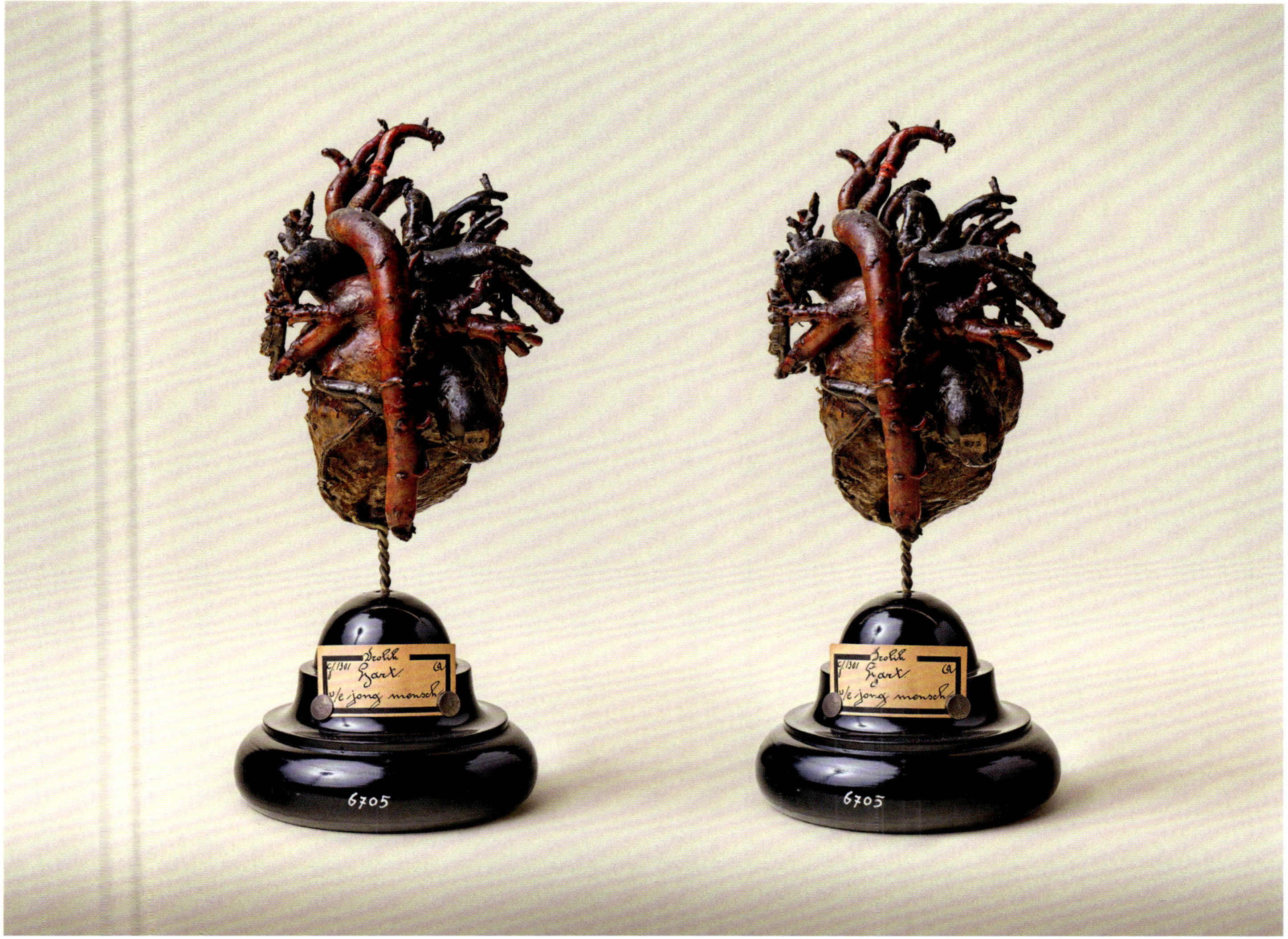
Vrolik
G/1301
Hart.
v/e jong mensch
6705
Vrolik
G/1301
Hart.
v/e jong mensch
6705

MALE GENITALIA AND URINARY TRACT

1890

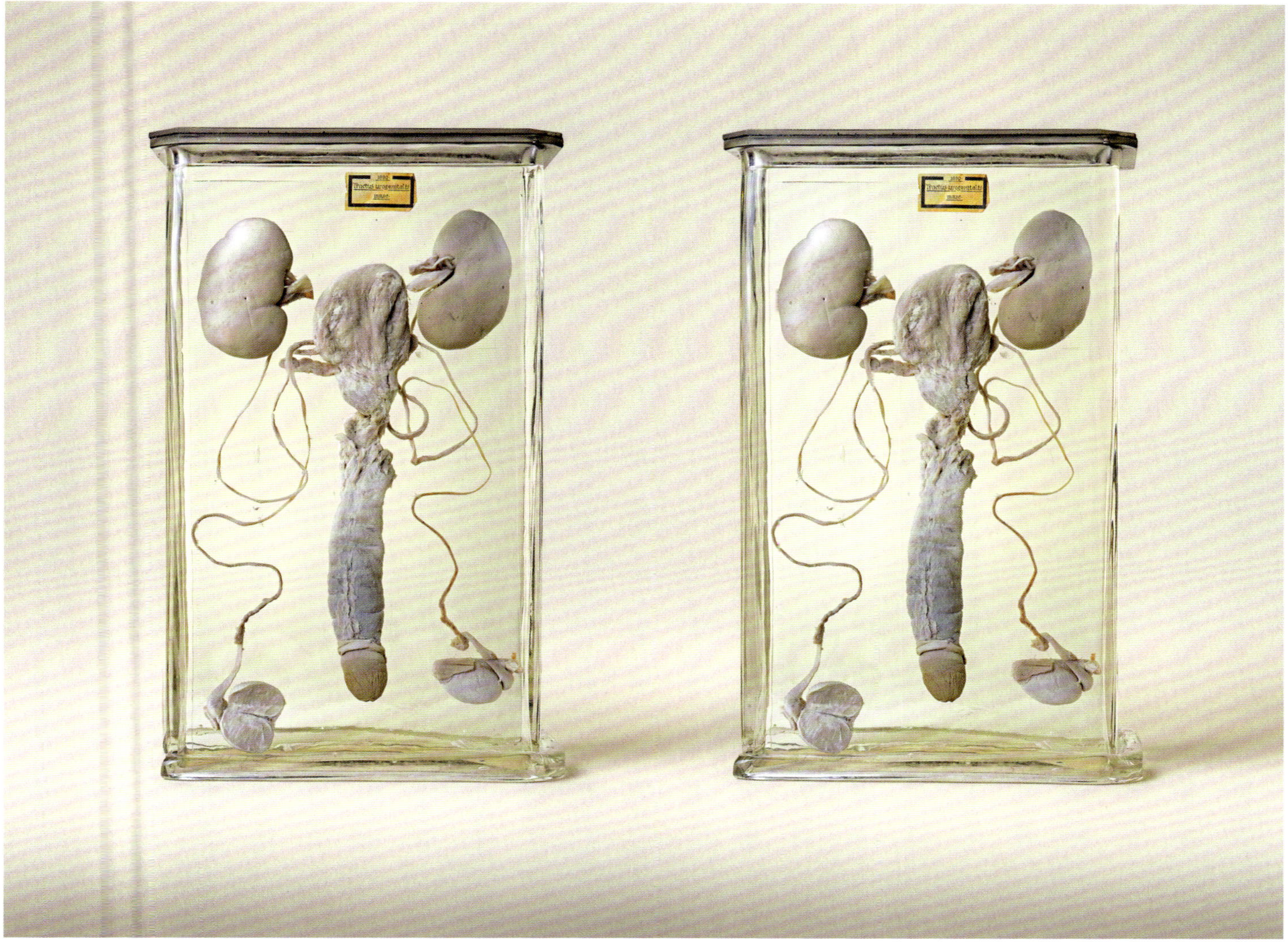
Tractus urogenitalis
Tractus urogenitalis

LONGITUDINAL SECTION OF A CHILD'S HEART

1892

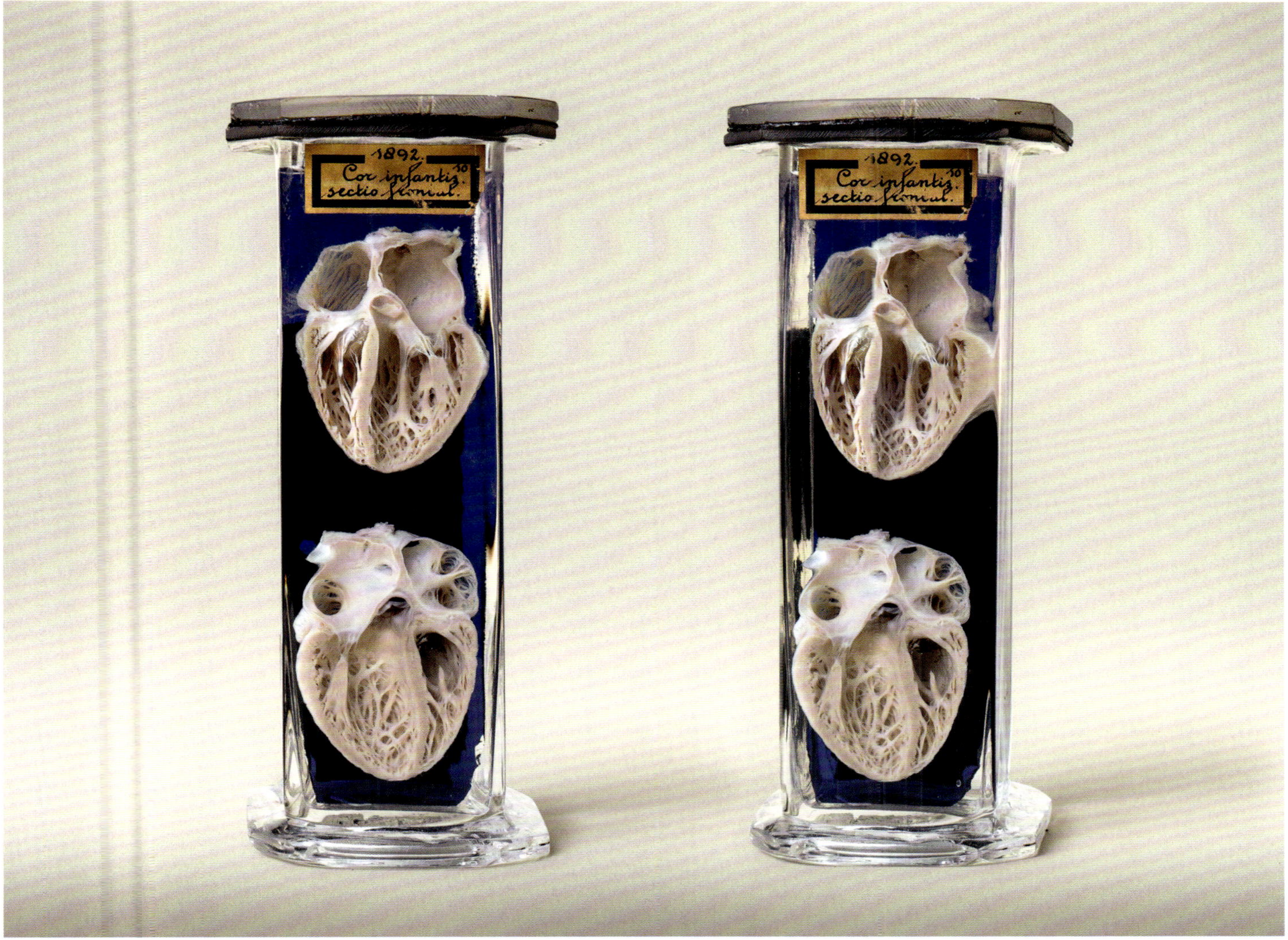
1892.
Cor infantis.
10
sectio frontal.

STOMACH OF A TWENTY-MONTH-OLD INFANT

1907

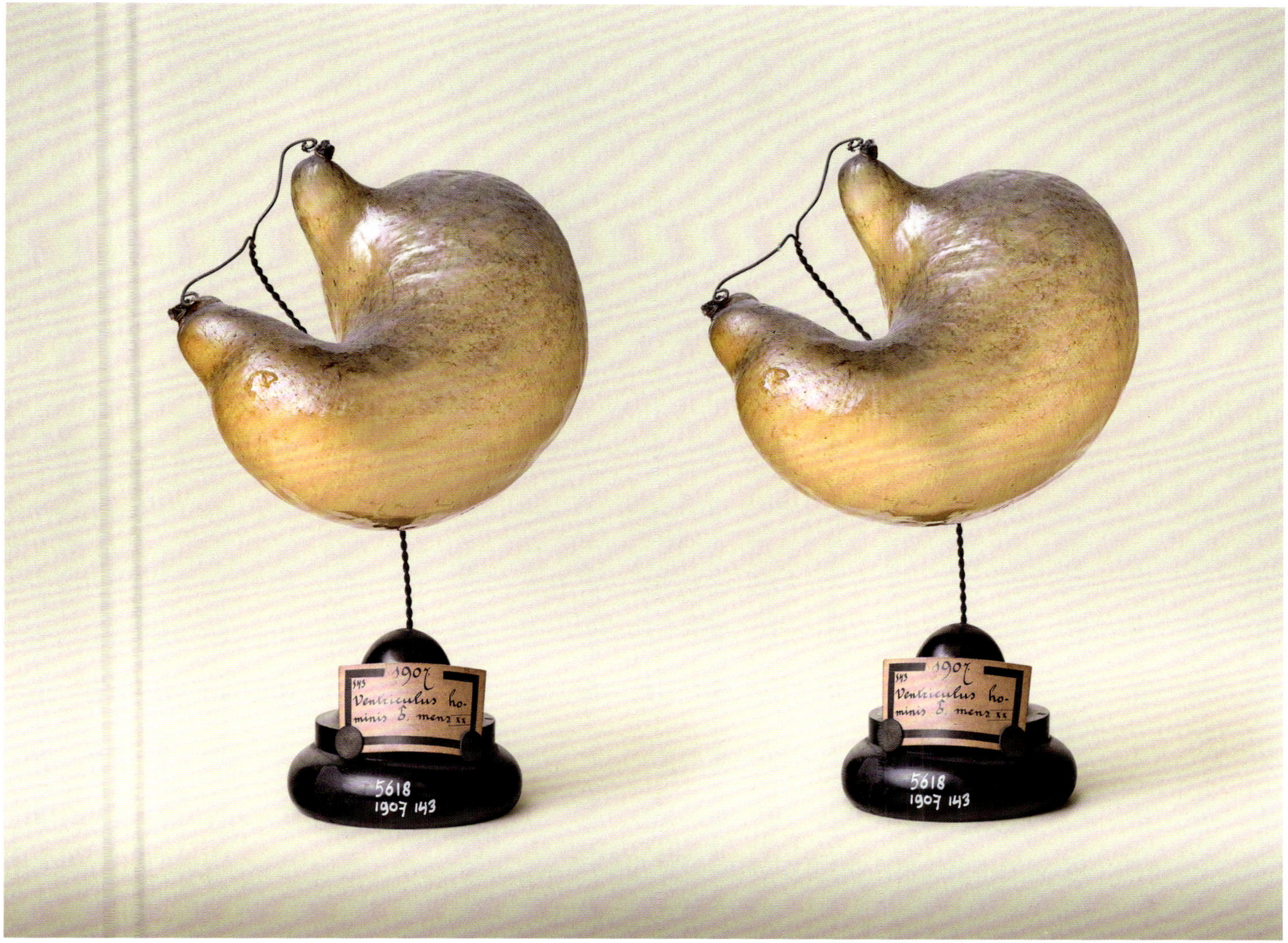
1907
Ventriculus ho-
5618
1907 143
1907
Ventriculus ho-
5618
1907 143

CROSS SECTION OF A FOETAL HEART
Late-nineteenth to early twentieth century

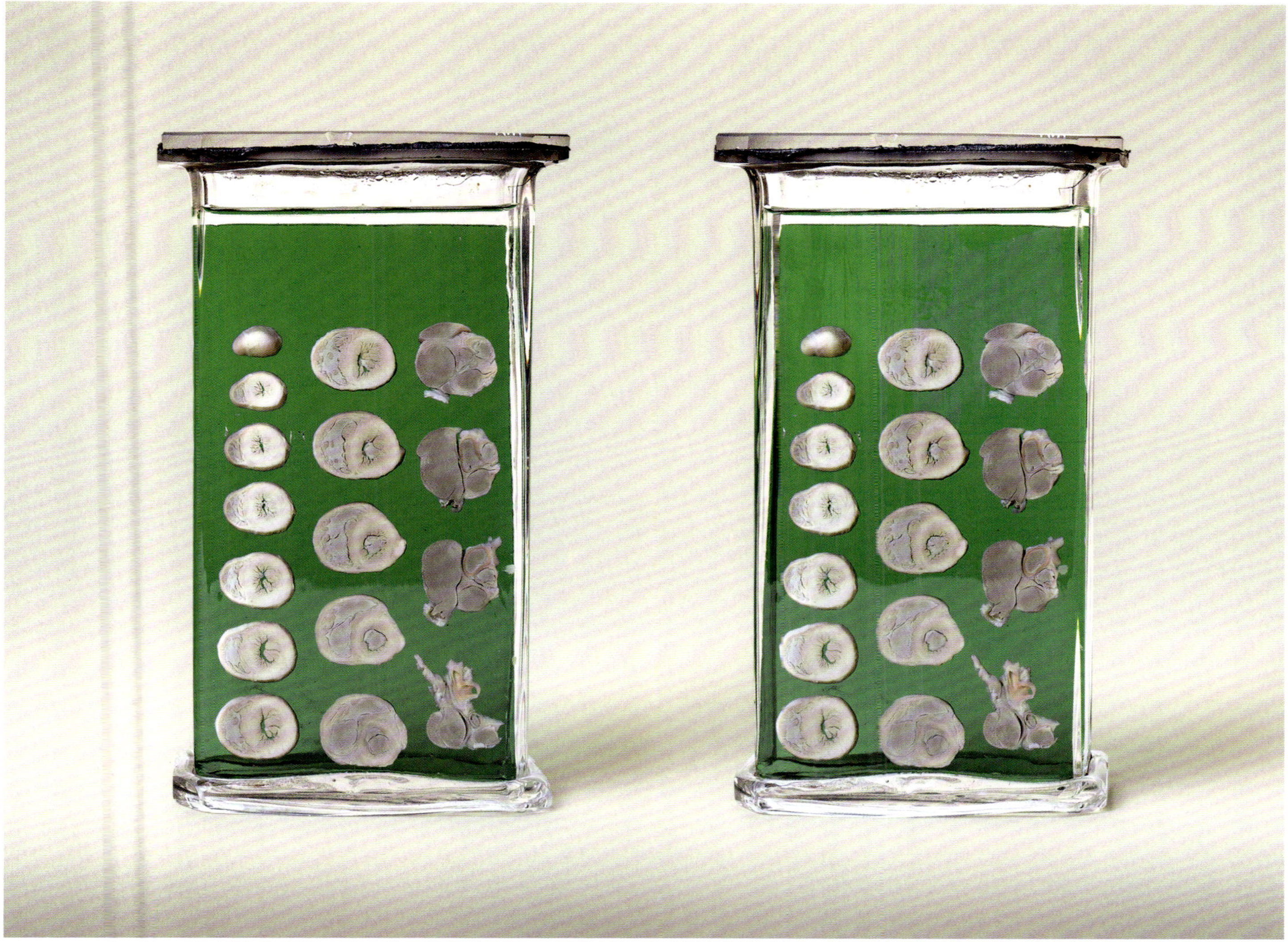

HEART AND AORTIC ARCH SHOWING A VARIATION IN THE BRANCHING OF THE LARGE BLOOD VESSELS

1902

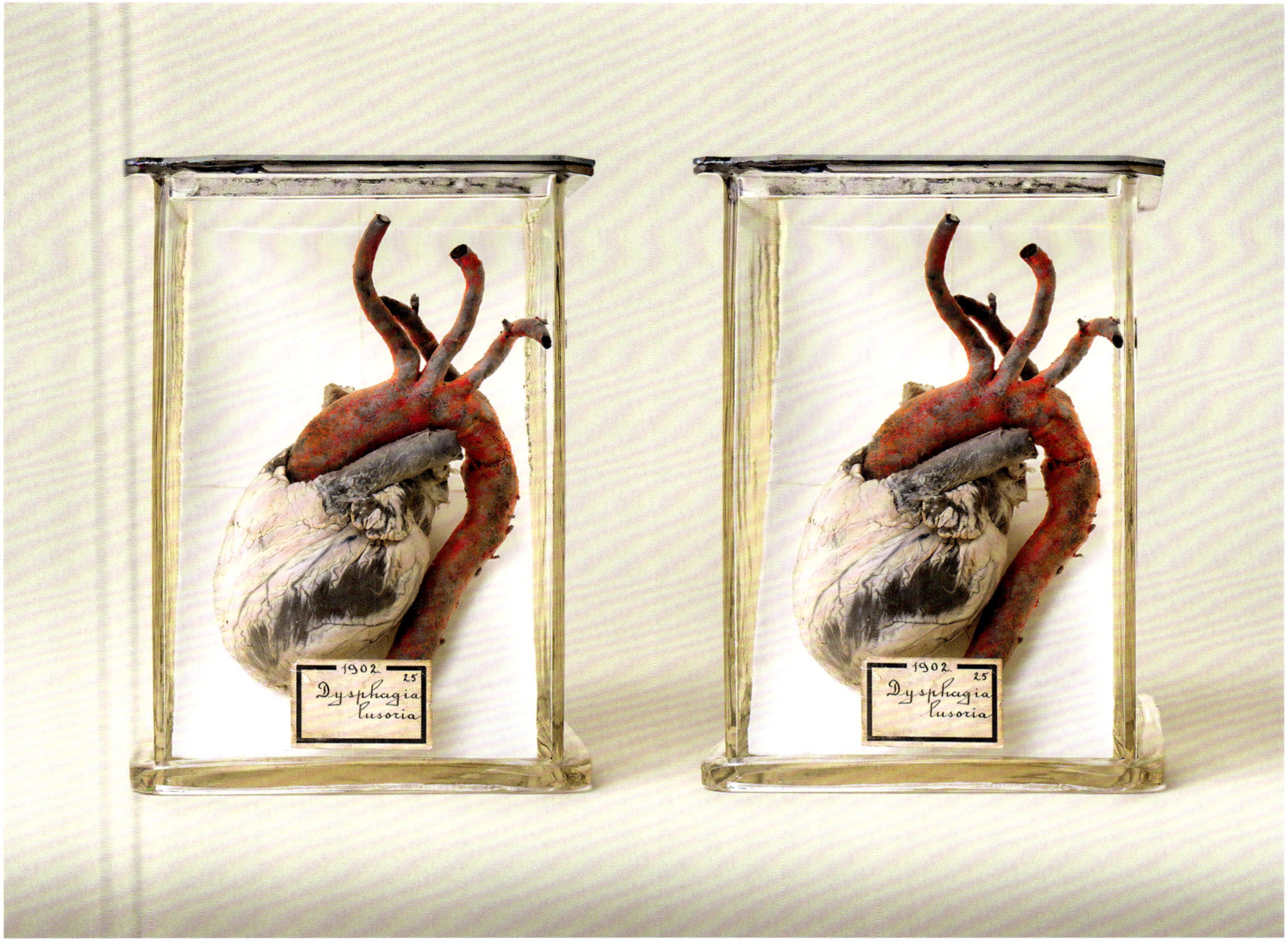

1902.
25
Dysphagia
lusoria
1902.
25
Dysphagia
lusoria

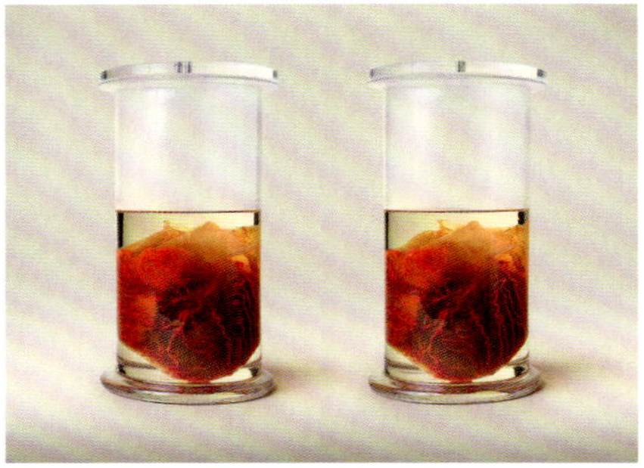

Adult heart showing the coronary arteries
Twentieth century
The heart has been conserved in a clearing agent, a fluid that makes the tissue translucent. The coronary arteries arise from the ascending aorta, and have been injected with a red dye. They are some of the most important blood vessels in the body, supplying the heart muscles with oxygenated blood. Blockage of these coronary arteries causes a myocardial infarction, commonly known as a heart attack.

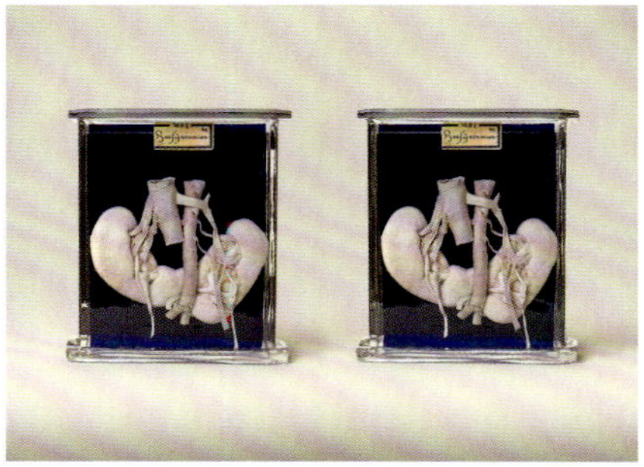

Horseshoe kidney
1884
A relatively common malformation, a horseshoe kidney originates during embryological development, when the lower ends of both kidneys fuse together. It is possible that someone who has a horseshoe kidney will only find out about it during a scan or an operation. Aside from the renal arteries and veins, the main vein and main artery of the body can be seen: the inferior vena cava (middle left) and the aorta (middle right), respectively.

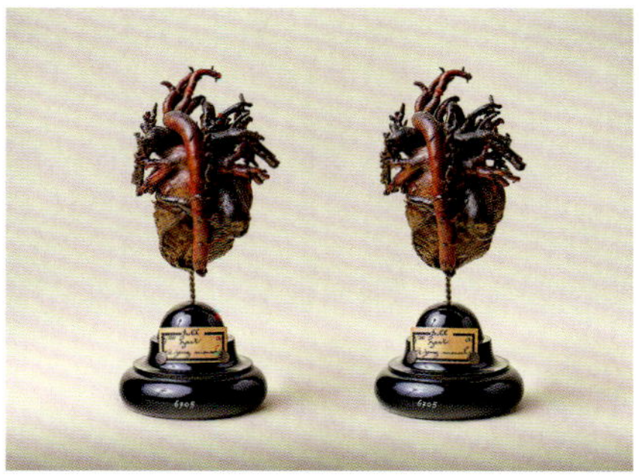

Heart of a young adult male
Before 1830
Showing the posterior aspect of the heart, this specimen has been dried and injected with red (signifying arteries) and dark blue (indicating veins) wax-like substances. The large vessel at the front is the descending aorta, the main artery of the body. The dark blue and red branches to the left and right are those of the pulmonary arteries and veins. The descending aorta passes down in front of the vertebral column.

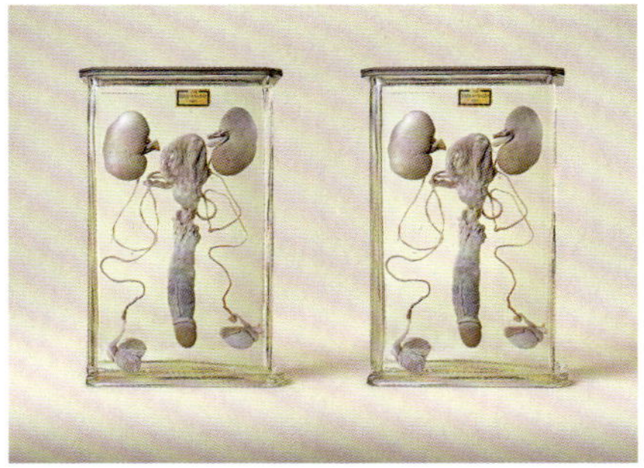

Male genitalia and urinary tract
1890
At the top left and right are the two kidneys, from which the two ureters deliver urine to the bladder (centre). From the bladder, urine is excreted through the urethra. This tube cannot be seen in this preparation because it runs through the prostate and the penis successively. About one third of the penis is located inside the body. The two testicles, one of which is in cross section, can be seen at the bottom of the specimen. Each of them is connected to the vas deferens, a long tube that transports mature sperm to the urethra.

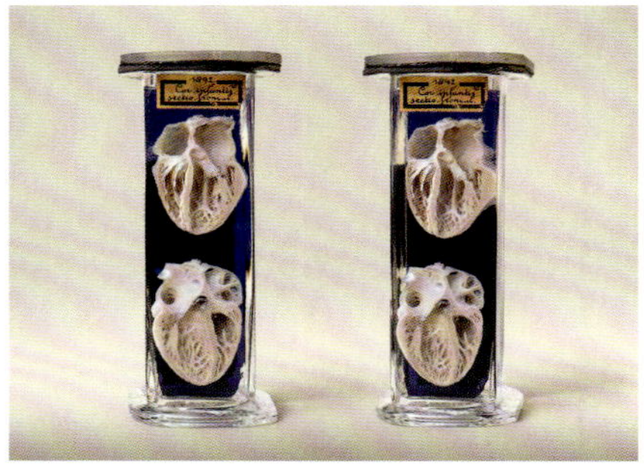

Longitudinal section of a child's heart
1892
The four chambers of the heart are clearly visible. In the bottom longitudinal section, the larger of the two chambers in the lower half of the heart is the left ventricle. The walls of both ventricles are more muscular than the walls of the atria, the two chambers located above the ventricles. In the top section, the left ventricle is located on the right-hand side. In this section, it is possible to see how some of the valves between the left atrium and ventricle are connected to muscles in the inner wall of the left ventricle. These muscles cause the valves to open to allow blood to pass from the atria into the ventricles.

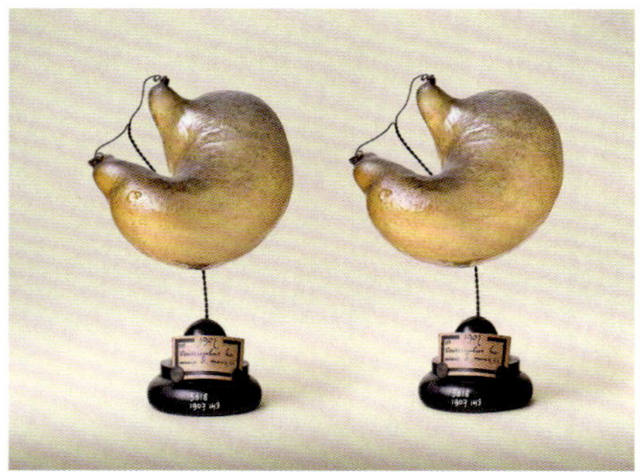

Stomach of a twenty-month-old infant
1907
This inflated and dried stomach of a 20-month-old boy shows its small inner curvature (left) and the larger outer curvature. When the stomach receives food it releases gastric acid and pepsin, an enzyme that breaks down proteins, creating partly digested food and digestive secretions called chyme. This process takes between half an hour and a few hours, depending on the quantity of food eaten. The chyme is then slowly but continuously pushed into the duodenum and the small intestine where nutrients are absorbed and further digestion takes place.

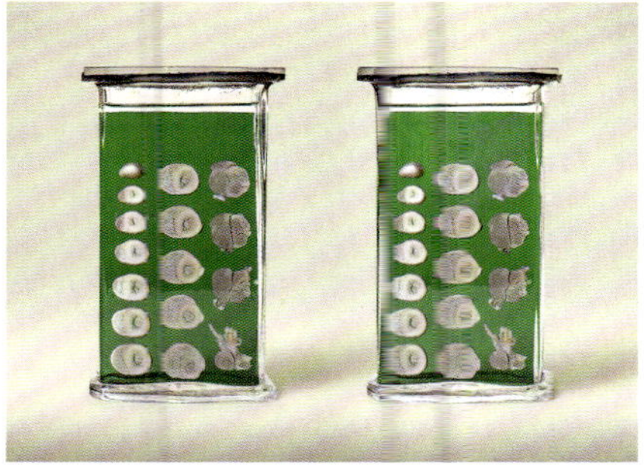

Cross section of a foetal heart
Late-nineteenth to early twentieth century
The first and second columns comprise cross sections of the ventricles, while the last column shows the atria. The specimen clearly demonstrates that the wall of the left ventricle (on the right of each of the cross sections) is much thicker and more muscular than the wall of the right ventricle. This is because the blood from the left ventricle needs to be pumped to the whole body, whereas that of the right ventricle is only pumped to the lungs. The right ventricle appears to be wrapped around the left ventricle.

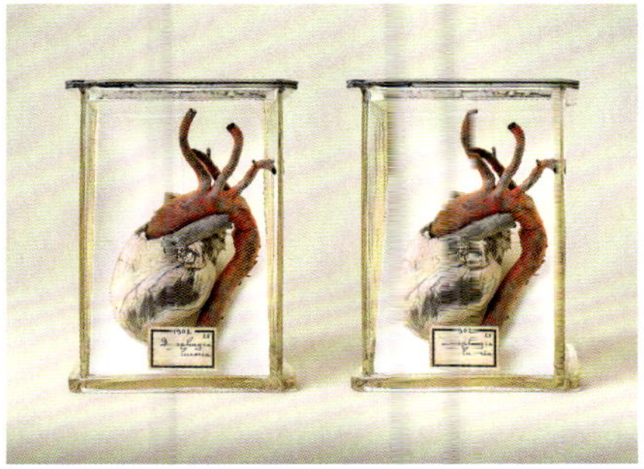

Heart and aortic arch showing a variation in the branching of the large blood vessels
1902
The aortic arch of the heart is dyed red. The four branching arteries of the head and arms are, from left to right: the right common carotid artery, the left common carotid artery, the left subclavian artery and, passing behind the other three, the right subclavian artery. Usually the right subclavian originates from a shared trunk with the right common carotid. However, here the right subclavian is located on the far right and has to pass over the oesophagus. Major arteries follow the pulse of the heart to transport blood, and the right subclavian may have obstructed the oesophagus, resulting in swallowing difficulties.

Bones

SKELETON OF AN ADULT MALE WITH ACHONDROPLASIA
Before 1830

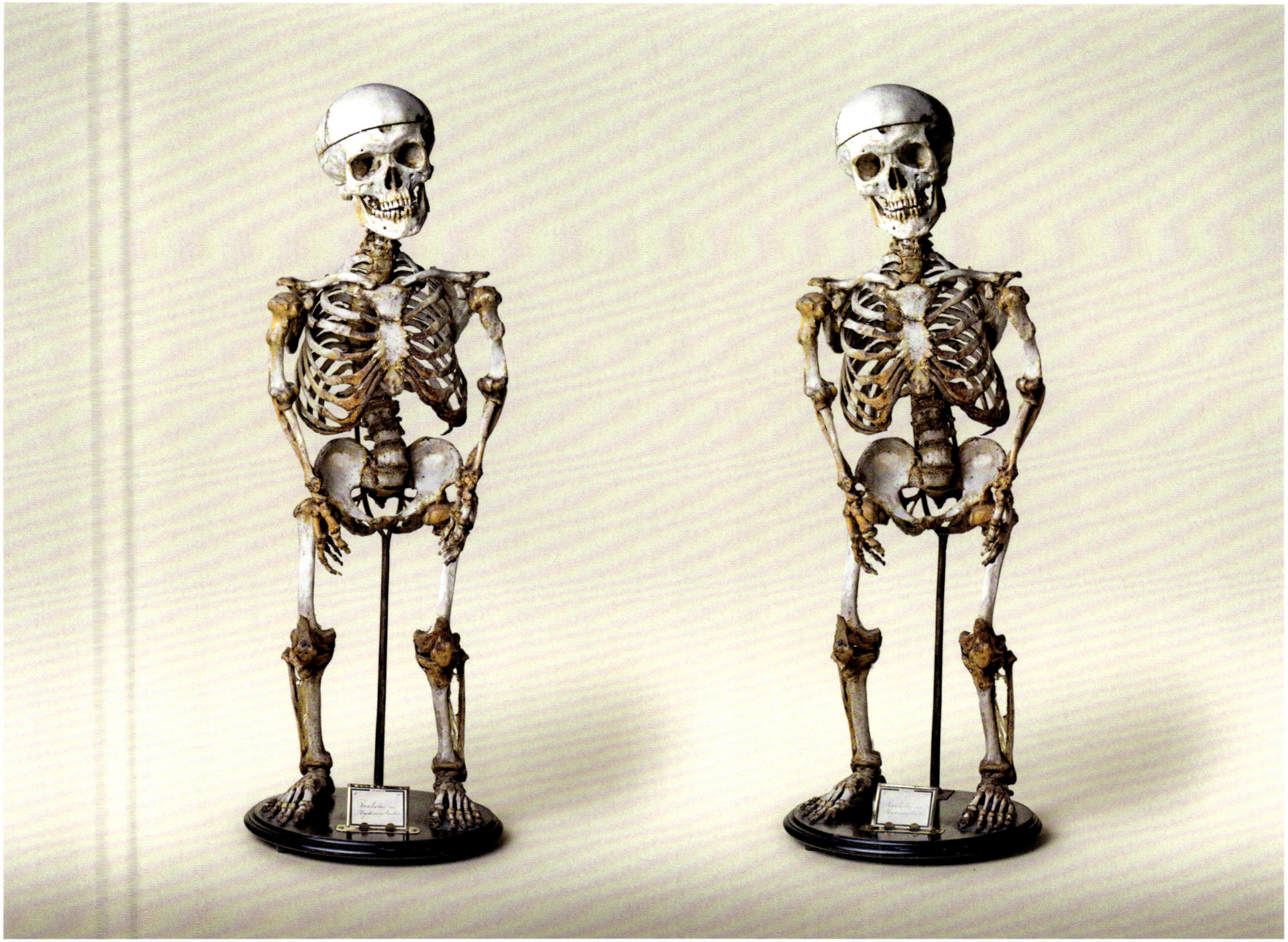

TRUNK WITH SCOLIOSIS OF THE VERTEBRAL COLUMN
Before 1830

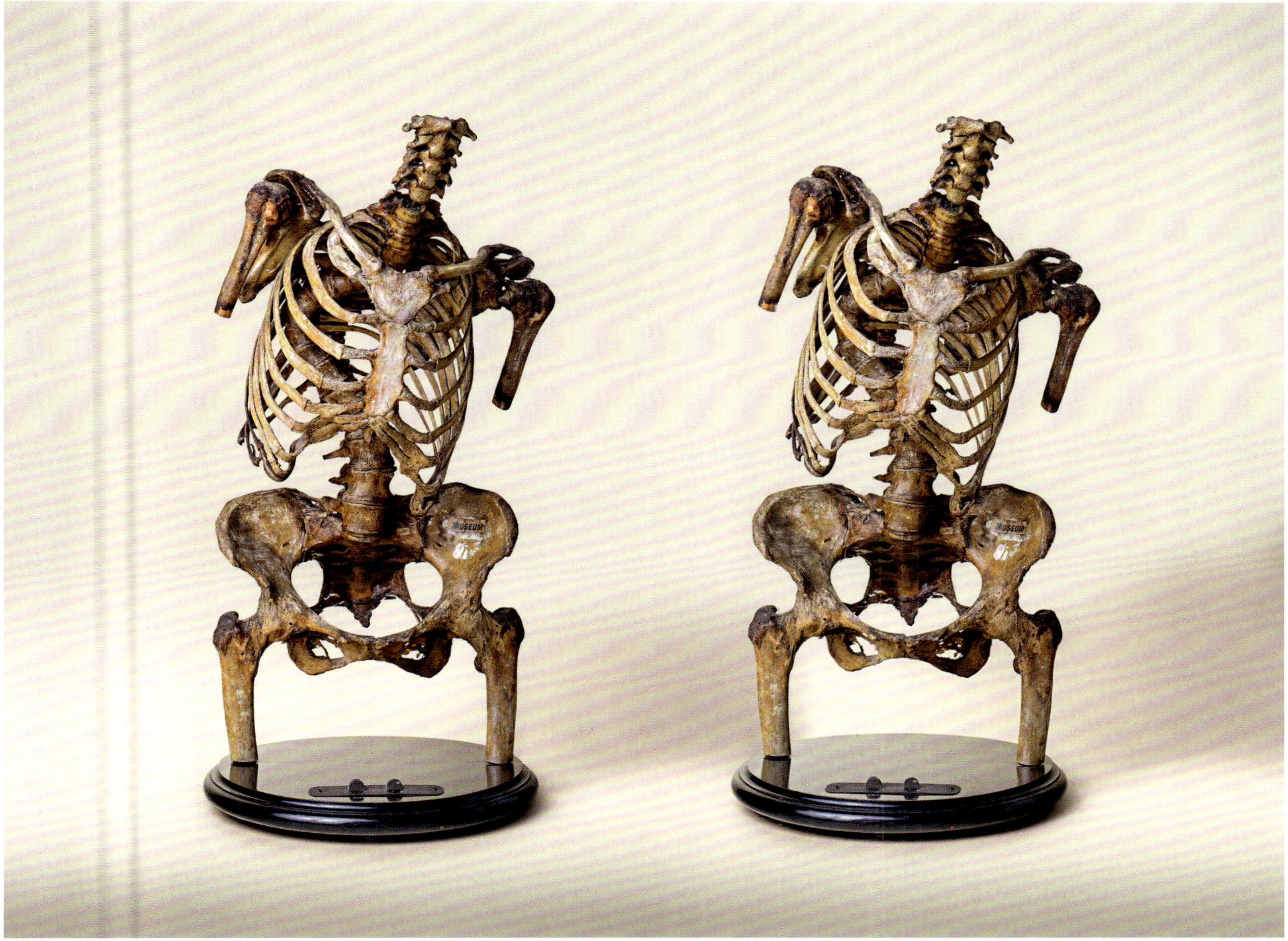

MAXILLA AND MANDIBLE WITH TEETH

1856–57

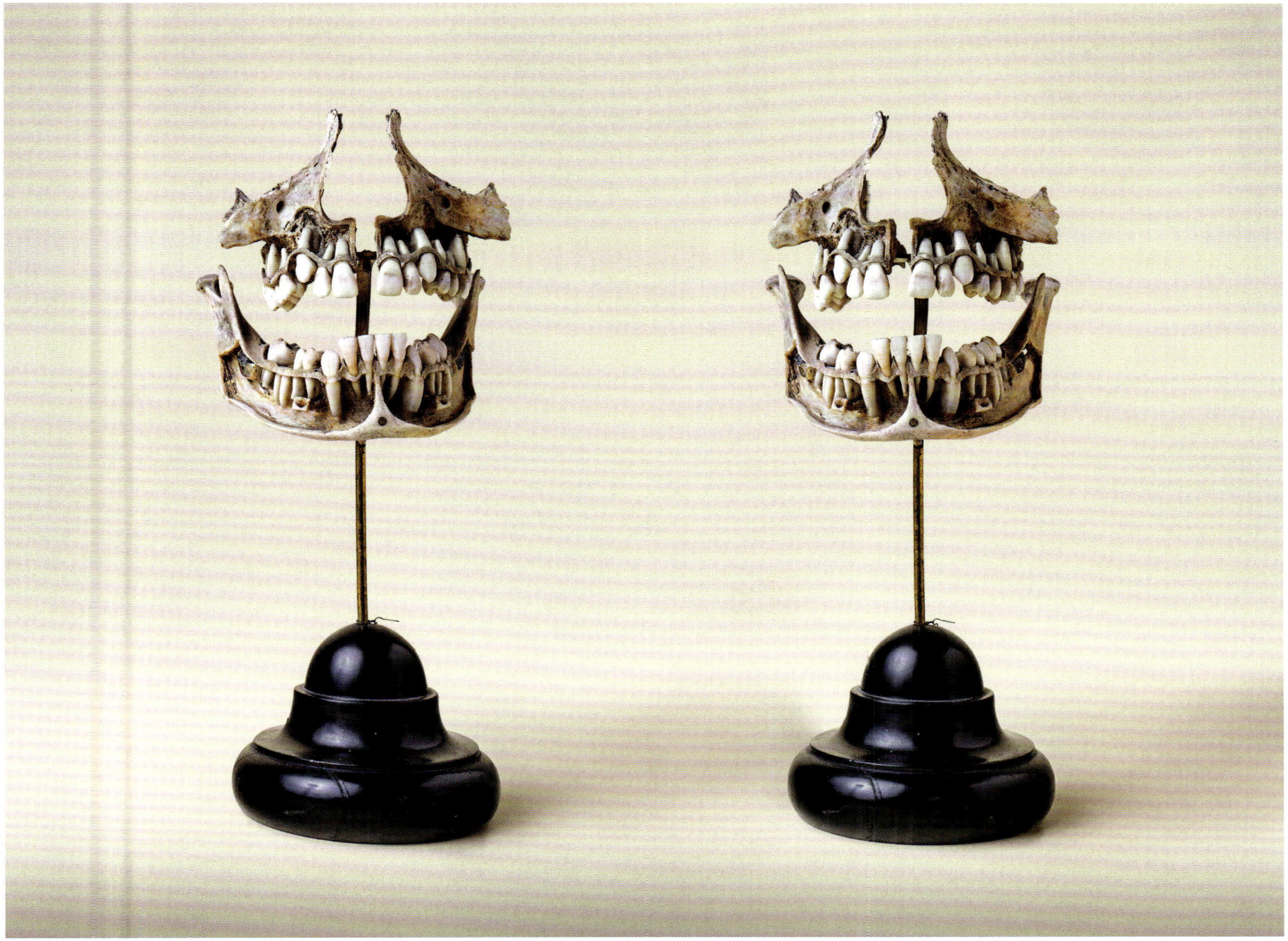

PELVIS OF ADULT FEMALE WITH RICKETS

1841

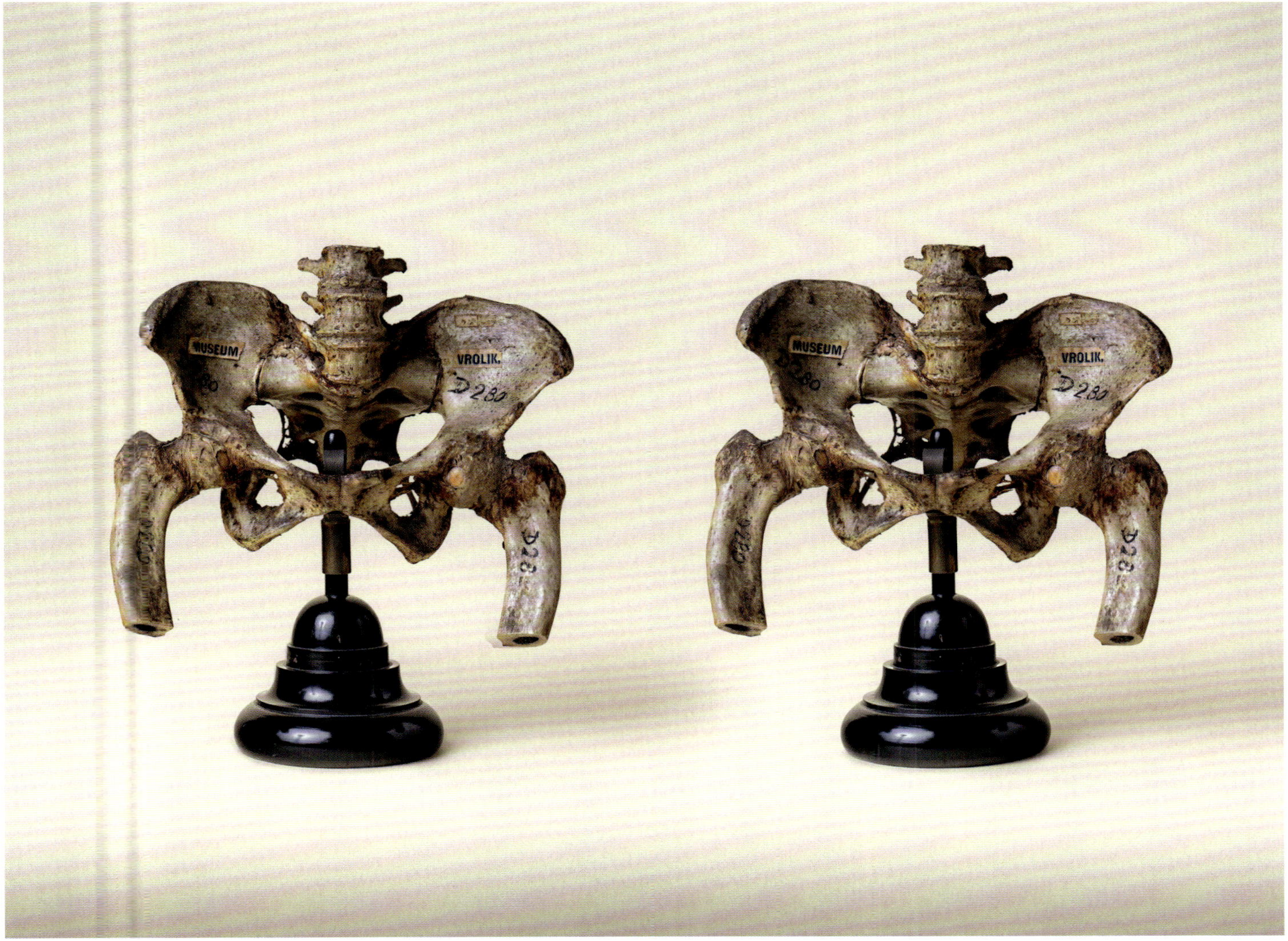
MUSEUM
VROLIK.
D280
MUSEUM
VROLIK.
D280

TRUNK OF A THREE-YEAR-OLD CHILD

Before 1830

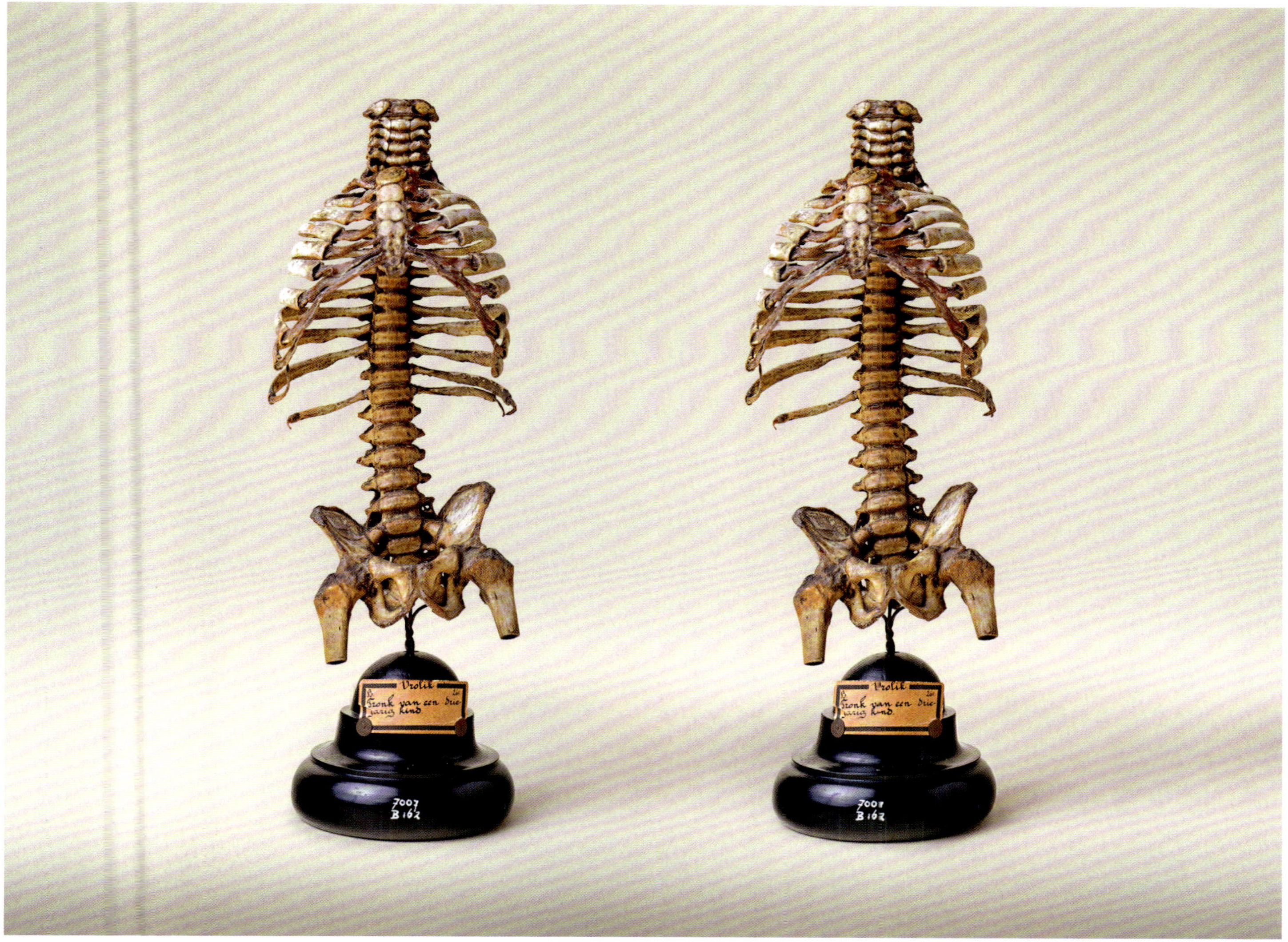
Vrolik
Tronk van een drie-
jarig kind
7007
B 162

SKELETON OF CONJOINED TWINS
c.1850

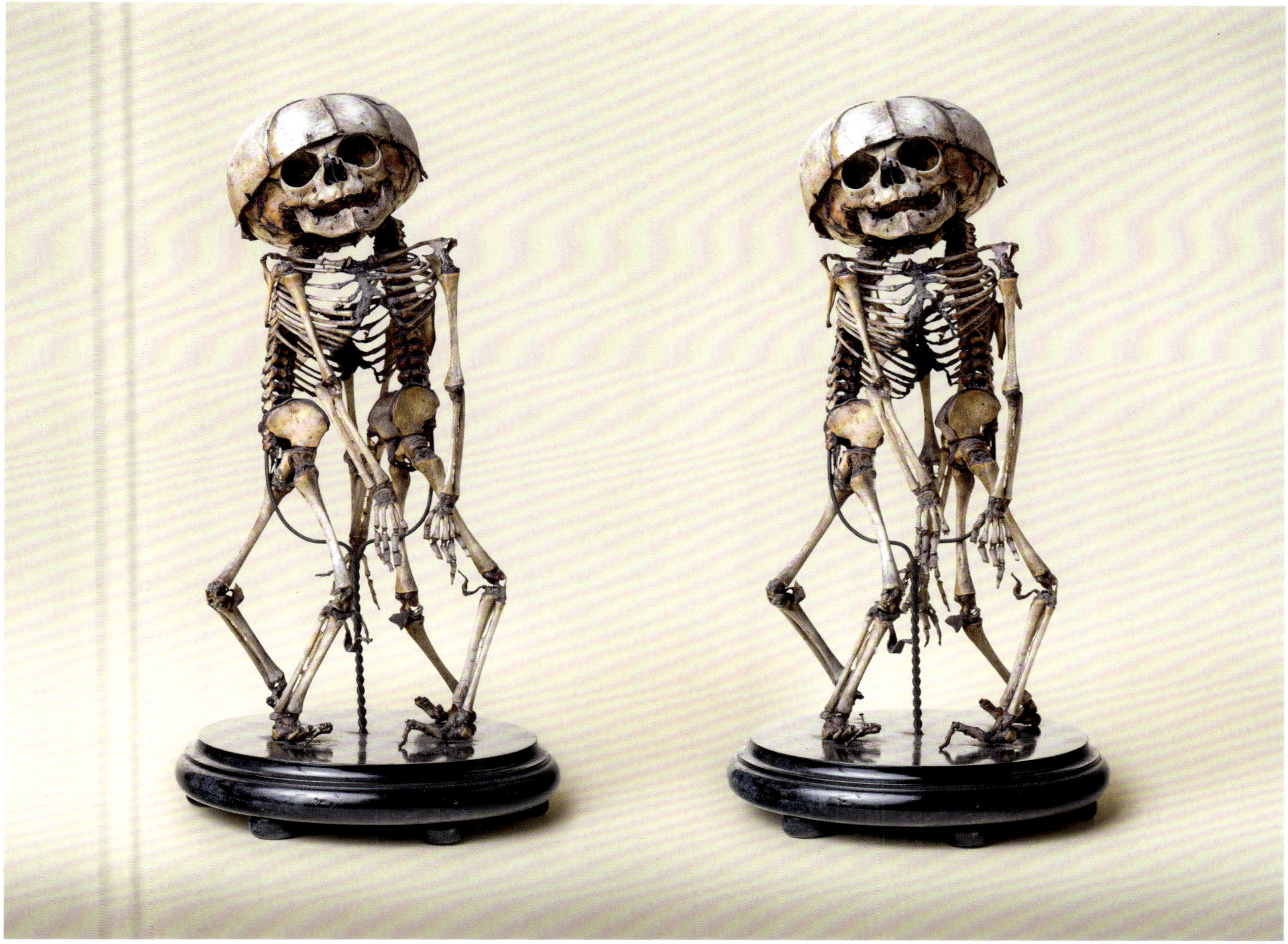

THORAX OF A NEWBORN INFANT
1889

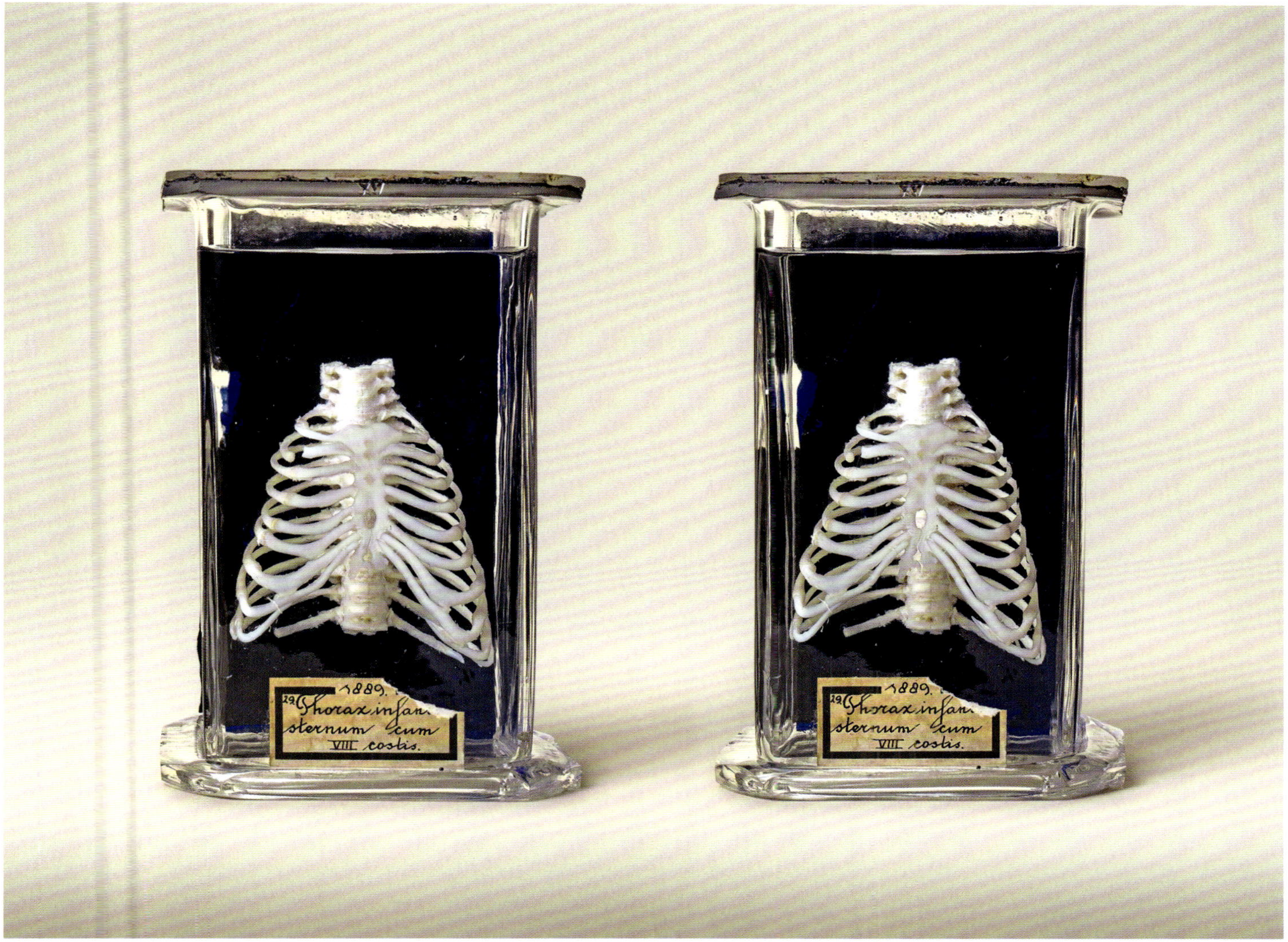
1889.
29. Thorax infan
sternum cum
VIII costis.
1889.
29. Thorax infan
sternum cum
VIII costis.

PELVIS SHOWING THE MAJOR BLOOD VESSELS OF THE LEG

1856–57

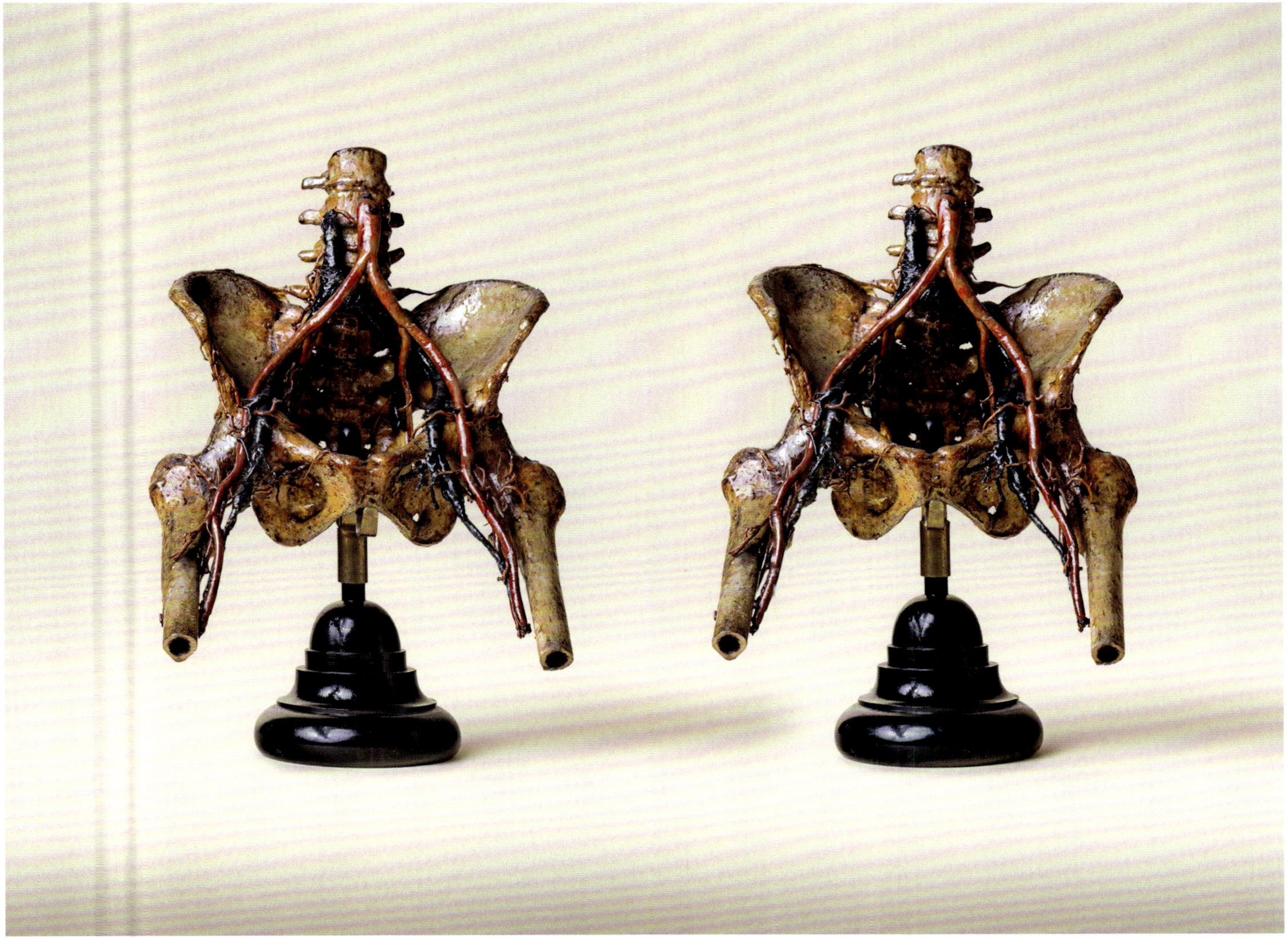

LEFT: SKELETON OF A THREE-YEAR-OLD BOY
c.1850

RIGHT: SKELETON OF A TWO-YEAR-OLD BOY
1856–57

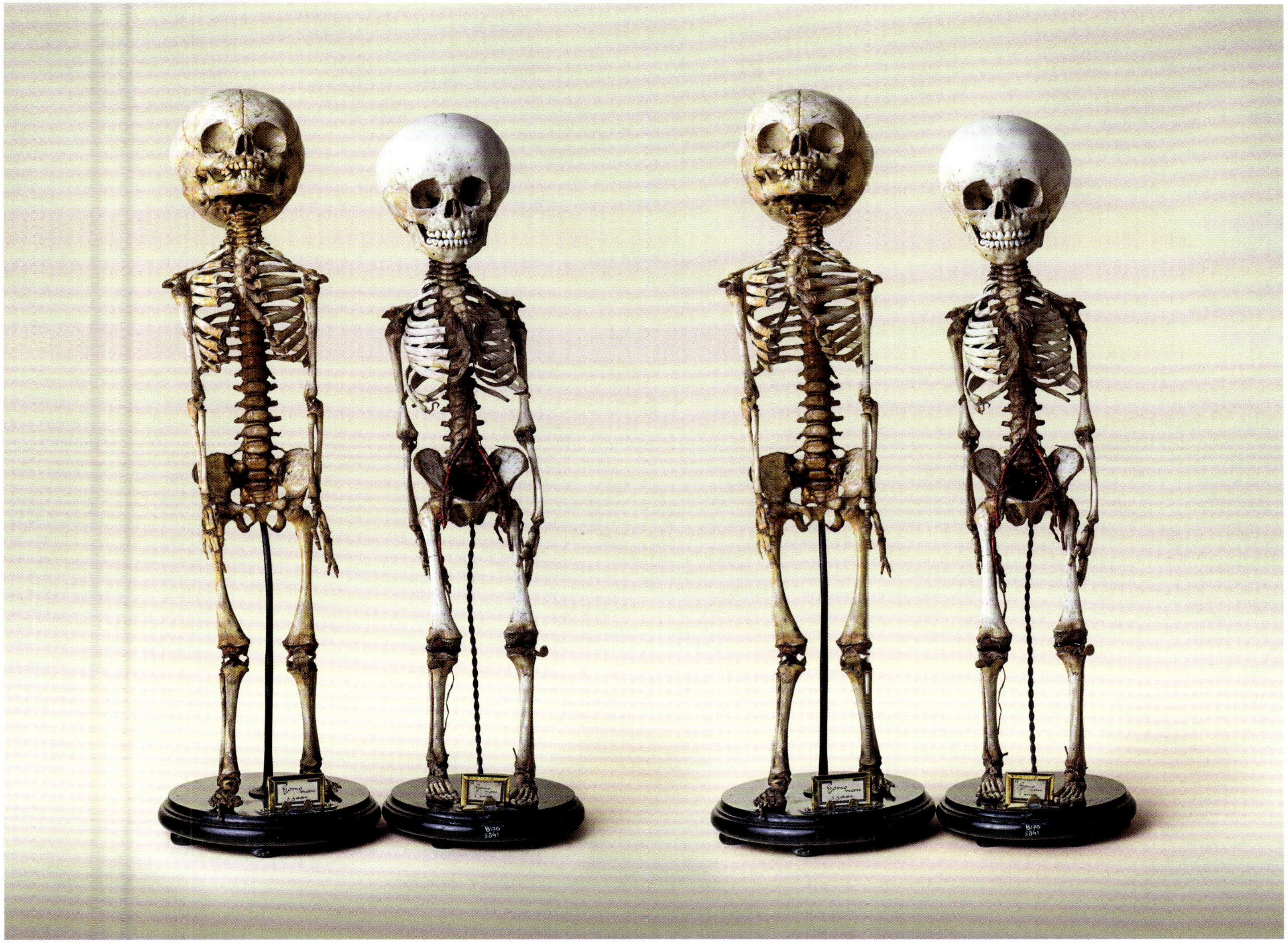

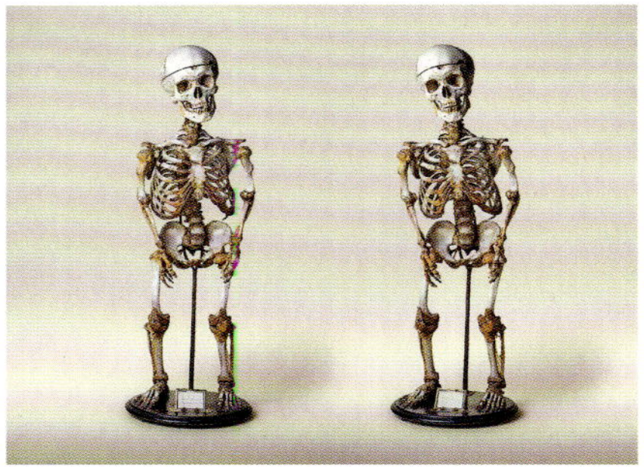

Skeleton of an adult male with achondroplasia
Before 1830
Achondroplasia, meaning 'underdeveloped cartilage', is a genetic disorder caused by mutations in the genes responsible for the development and ossification of the cartilage, resulting in a substantial lack of longitudinal growth of the bones. People with achondroplasia, or dwarfism, usually live a normal life. In this case, the skeleton belonged to a man who worked as a school teacher and died at the age of 60. In the nineteenth century, achondroplasia and other forms of skeletal dysplasia were not yet identified as separate disorders, and Willem Vrolik believed that it was a form of congenital rickets.

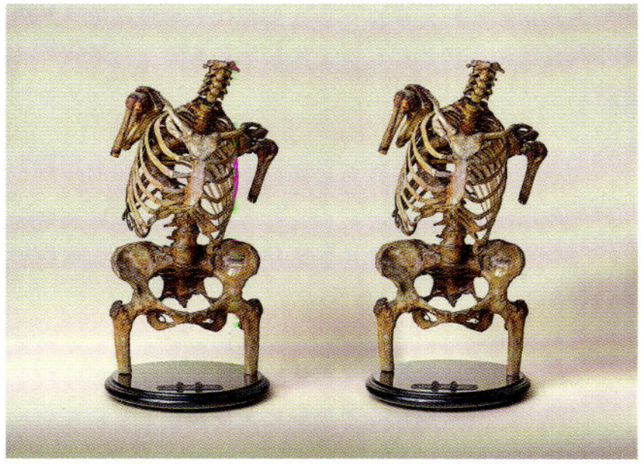

Trunk with scoliosis of the vertebral column
Before 1830
The vertebral column shows scoliosis, which is characterized by a sideways bend of the spine (not by a bend to the front, which is called kyphosis). In scoliosis, the curve is usually C- or S-shaped because the body attempts to correct the posture. Over time the curve may worsen, especially if it is not helped with such orthoses as a brace, although these were rarely available in the nineteenth century. This may explain why the Vrolik collection counts so many examples of severe scoliosis. In most cases, the cause of scoliosis is not clear.

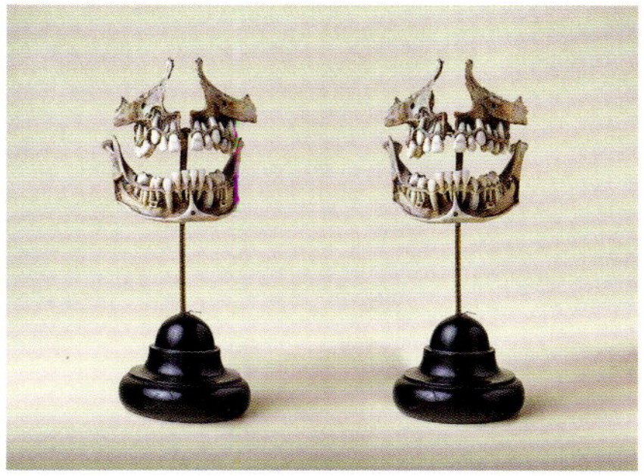

Maxilla and mandible with teeth
1856–57
The outer layer of bone has been removed to show the roots of the permanent teeth. Most deciduous teeth have been replaced by permanent teeth by the age of 13, the last of which to erupt are the canines. The third molars, better known as wisdom teeth, appear between the ages of 17 and 25.

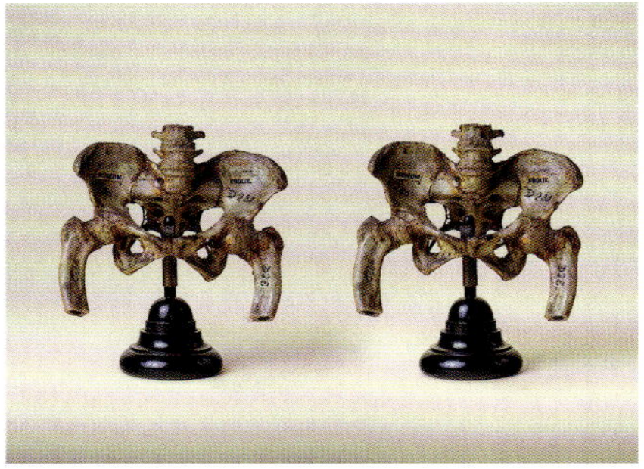

Pelvis of adult female with rickets
1841
Dying shortly after labour, this lady had a narrow pelvis, which turned out to be disastrous for both her and her child. Due to rickets, her pelvis was flattened and the pelvic inlet restricted. When contractions started, the infant got stuck inside the pelvis, and by the time the obstetrician arrived the child had died. A caesarian section was performed, and, consequently, the mother died a few days later of peritonitis and puerperal fever. In the nineteenth century, operations of the abdomen such as a caesarian section, carried high risk of death due to the lack of knowledge about contamination and antisepsis.

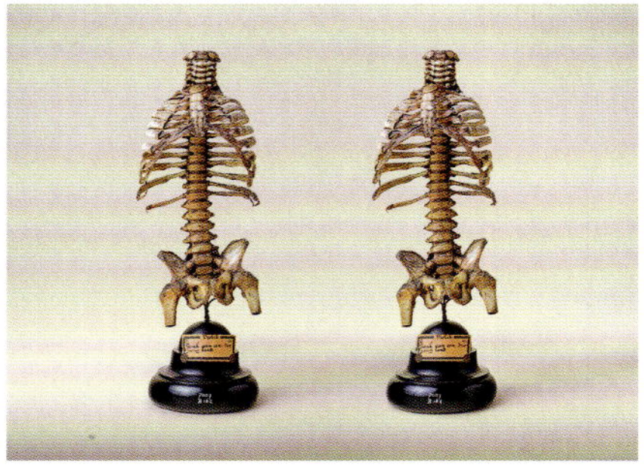

Trunk of a three-year-old child
Before 1830
The ossification (bone formation) of the skeleton is in full progress: the sternum still consists of several separate parts, and the pelvis still comprises several elements with thin layers of cartilage between them. In these layers of cartilage, new bone will form, allowing the skeleton to grow. By around the age of 20, the pelvis (and most other bones) will be fully grown and the layers of cartilage will disappear.

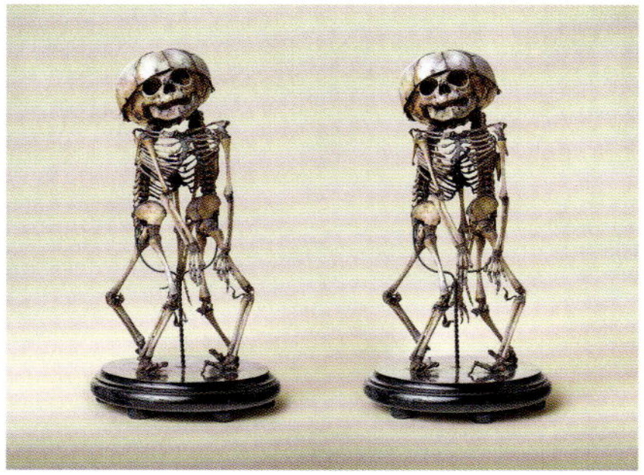

Skeleton of conjoined twins
c.1850
The twins are fused together ventrally from head to abdomen. The skull and the chest appear normal, but, in fact, are composed of half the face and chest of the left individual and half the face and chest of the right individual, fused together. Conjoined twins of this type are stillborn because their fused brains and hearts do not function. In order to study the morphology of the conjoined brain, anatomist Willem Vrolik opened the cranial cavity.

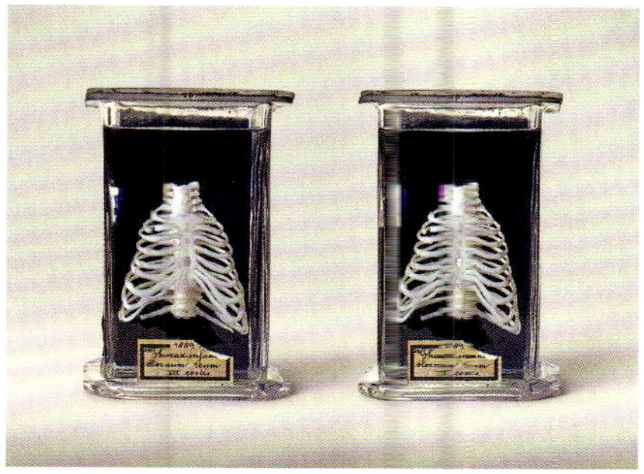

Thorax of a newborn infant
1889
Two anatomical variations are shown on the thorax. The first is the little hole in the sternum; the second is the number of fixed ribs. During embryological development, the thoracic wall closes from the sides because connective tissue from both halves of the primordial sternum grow together and fuse. Sometimes this process remains incomplete, hence the small hole seen here. The rib cage is made up of twenty-four bones arranged in twelve pairs. Fixed ribs are joined directly to the sternum by costal cartilage, and usually number seven pairs. In this case, there are eight pairs of fixed ribs.

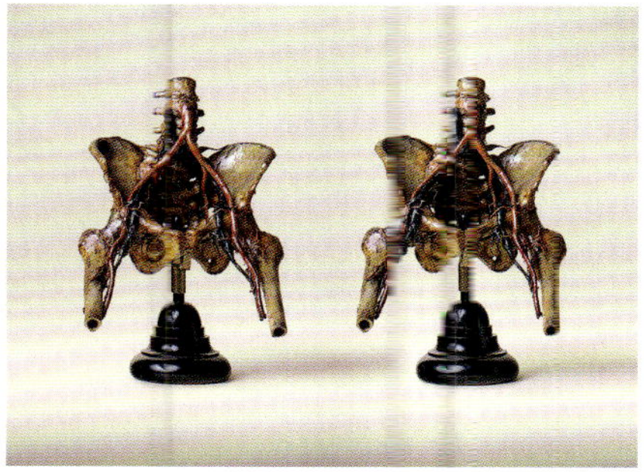

Pelvis showing the major blood vessels of the leg
1856–57
Injected with a coloured wax-like substance, the arteries that carry the oxygen-rich blood from the heart to the rest of the body are shown in red. The veins that bring the blood back to the heart are filled with dark blue wax. The aorta, the main artery of the body that passes down the abdominal cavity, splits within the pelvis into left and right common iliac arteries. In turn, these common iliac arteries divide into internal and external iliac arteries. The internal ones pass to the pelvic organs and bring blood to the bladder, rectum and (in women) uterus. The external ones become the femoral arteries of the upper legs.

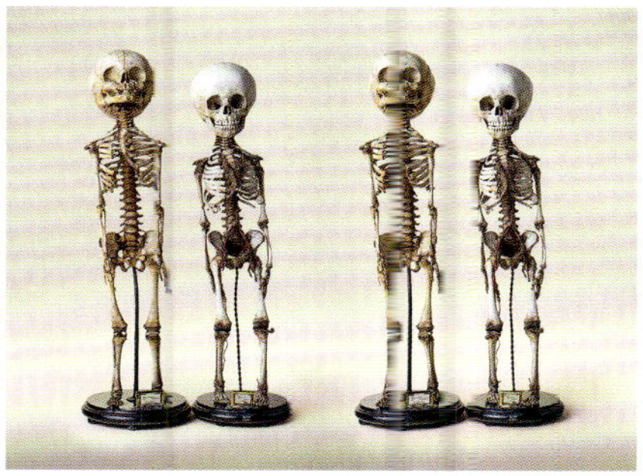

Left: Skeleton of a three-year-old boy
c.1850
Right: Skeleton of a two-year-old boy
1856–57
Remarkably the deciduous teeth of the three-year-old boy have not all appeared, whereas they have in the two-year-old. It is also striking that the two hardly differ in height. This is in part due to the natural variation in growth and development in children, but may also be the result of stunted growth and development due to disease and malnutrition. The skeletons are from the nineteenth century when the child mortality rate in Europe was high. In the right skeleton, the major arteries and veins of the leg are visible, having been injected with coloured wax.

ARTIST BIOGRAPHY

Jim Naughten
b.1969

Education
Arts Institute of Bournemouth 1990–92

Exhibitions
Manchester Museum (August 2016)
'Artijuli' ('Art in July'), Krakeslottet, Norway (July 2016)
Museum of Photographic Arts, San Diego (June–October 2016)
KK Outlet, London (May 2016)
Powell-Cotton Museum, Margate (April–September 2016)
Klompching Gallery, New York (April–May 2016)
AIPAD, New York (April 2016)
Horniman Museum, London (September 2015–January 2016)
Christie's, London, 'Out of the Ordinary' (August–September 2015)
Royal Academy of Arts, London, 'Summer Exhibition 2015' (June–August 2015)
Couleur Café Festival, Brussels (June 2015)
Klompching Gallery, New York (September–October 2014)
Noorderlicht Gallery, Groningen (July–August 2014)
ArtiJuly, Krakeslottet, Norway (July 2014)
Portrait(s) Festival, Vichy (July 2014)
Toronto Photography Festival (May 2014)
AIPAD, New York (April 2014)
Artist Residence, Krakeslottet, Norway (March 2014)
Knokke-Heist Photo Festival, Belgium (March 2014)
Honolulu Museum of Art (December 2013–June 2014)
Imperial War Museum North, Manchester, 'Catalyst: Contemporary Art and War' (October 2013–November 2014)
MIA Gallery, Seattle (August 2013)
Annenberg Space Gallery, Los Angeles (May 2013)
Klompching Gallery, New York (March 2013)
Margaret Street Gallery, London (March 2013)
Royal Shakespeare Company, London (November 2012–March 2013)
Imperial War Museum, London (June–December 2012)
Royal Academy of Arts, London, 'Summer Exhibition 2012' (May–August 2012)
Quay No.1 Vevey, Switzerland (May–June 2012)
Foto8 Gallery (2012)
Gallery Huit, Malaysia (2012)
Photographers' Gallery, Victoria Park, London (August 2011)
Gallery Huit, Arles (July 2011)
Barbara Frigerio Contemporary Art, Milan (2010)
Klompching Gallery, New York (2010)
HotShoe Gallery, London (2010)
National Portrait Gallery, London (2009)
AOP Gallery, London (2009)
Coningsby Gallery, London (2001)
Coningsby Gallery, London (2000)
ICA Gallery, London (1995)

Books (monographs)
Animal Kingdom (Hoop Editions and Prestel, 2016)
Conflict and Costume (Merrell, 2013)
Re-enactors (HotShoe, 2009)

Books (group)
Haute Africa (Lannoo, 2014)
Afterwards (Thames & Hudson, 2011)

FURTHER READING

Alberti, Samuel J. M. M., *Morbid Curiosities: Medical Museums in Nineteenth-Century Britain*, Oxford University Press, 2011

Alberti, Samuel J. M. M., and Elizabeth Hallam, eds., *Medical Museums: Past, Present, Future*, Royal College of Surgeons of England, 2013

Barnett, Richard, *Crucial Interventions: An Illustrated Treatise on the Principles & Practice of Nineteenth-Century Surgery*, Thames & Hudson, 2015

Barnett, Richard, *The Sick Rose: Or, Disease and the Art of Medical Illustration*, Thames & Hudson, 2014

Bourgery, J. M., *The Complete Atlas of Human Anatomy and Surgery*, Taschen, 2015

College of Physicians of Philadelphia, *Mütter Museum: Historic Medical Photographs*, Blast Books, 2007

Degueurce, Christophe, *Fragonard Museum: The Écorchés. The Anatomical Masterworks of Honoré Fragonard*, Blast Books, 2011

Ebenstein, Johanna, *The Morbid Anatomy Anthology*, Morbid Anatomy, 2014

Evans, Emily, *Anatomy in Black*, Lotus Publishing, 2016

Gambino, Paul, *Morbid Curiosities: Collections of the Uncommon and the Bizarre*, Laurence King Publishing, 2016

Gray, Henry, *Gray's Anatomy* (Leatherbound Classic Collection), Barnes & Noble, 2011

Hausner, Ernst, *The Fool's Tower: the Federal Pathological-anatomical Museum at the Old General Hospital in Vienna*, Edition Hausner, 1998

Hudspeth, E. B., *The Resurrectionist: The Lost Work and Writings of Dr Spencer Black*, Quirk Books, 2013

Hunter, Michael, *Boyle: Between God and Science*, Yale University Press, 2010

Kooijmans, Luuc, *Death Defied: the Anatomy Lessons of Frederik Ruysch*, trans. Diane Webb, Brill, 2011

Moore, Wendy, *The Knife Man: The Extraordinary Life and Times of John Hunter, Father of Modern Surgery*, Broadway Books, 2005

Pellerin, Denis and Brian May, *The Poor Man's Picture Gallery*, Carlton Books, 2014

Rifkin, Benjamin A., *Human Anatomy: Depicting the Body from the Renaissance to Today*, Thames & Hudson, 2011

Rooy, Laurens de, and Hans van den Bogaard *Forces of Form: The Vrolik Museum*, Amsterdam University Press, 2009

Ruysch, Frederik, *The Celebrated Dr Frederic Ruysch's Practical Observations in Surgery and Midwifry*, Nabu Press, 2014

Sappol, Michael, *Dream Anatomy*, National Library of Medicine, 2006

Virchow, Rudolf, *Cellular Pathology as Based Upon Physiological and Pathological Histology* (Classic Reprint), Forgotten Books, 2012

COLLECTION CREDITS

All collections: Museum Vrolik AMC

Sub-collections at Museum Vrolik

Collection Willem Berlin (1825–1902): page 51
Collection Louis Bolk (1866–1930): pages 23, 27, 33, 43, 47, 49, 53, 55, 57, 65, 67, 73, 85, 91, 111, 115
Collection of the Museum of Anatomical Theatre: pages 71, 79, 125, 135, 137
Collection unknown: pages 19, 29, 95, 101, 113
Collection Max Fürbringer (1846–1920): pages 45, 69, 87, 103
Collection Georg Ruge (1852–1919): front cover and pages 81, 89, 93, 107, 109, 133
Collection Gerard (1775–1859) and Willem (1801–63) Vrolik: pages 21, 35, 41, 63, 83, 105, 121, 123, 127, 129, 131
Collection Martin Woerdeman (1892–1990): pages 25, 31

PHOTOGRAPHY CREDITS

INDEX

First published in 2017 by Prestel Verlag
in association with Hoop Editions, an imprint of Hoop Design Limited
www.hoopdesign.co.uk
Tel. +44 (0)1323 832 137

Prestel Verlag, Munich
A member of Verlagsgruppe Random House GmbH

Prestel Verlag
Neumarkter Strasse 28
81673 Munich
Tel. +49 (0)89 4136-0
Fax +49 (0)89 4136-2335
www.prestel.de

Prestel Publishing Ltd.
14–17 Wells Street
London W1T 3PD
Tel. +44 (0)20 7323 5004
Fax +44 (0)20 7323 0271

Prestel Publishing
900 Broadway, Suite 603
New York, NY 10003
Tel. +1 (212) 995-2720
Fax +1 (212) 995-2733
www.prestel.com

Library of Congress Control Number is available. British Library Cataloguing-in-Publication Data: a catalogue record for this book is available from the British Library; Deutsche Nationalbibliothek holds a record of this publication in the Deutsche Nationalbibliografie; detailed bibliographical data can be found under: http://dnb.d-nb.de

Prestel books are available worldwide. Please contact your nearest bookseller or one of the above addresses for information concerning your local distributor.

Created and produced for Prestel Verlag by Hoop Editions
Editorial direction: Catherine Hooper
Graphic design and production: Adam Hooper
Proofreader: Jenny Wilson
Separations: Reproline Mediateam
Printing and binding: C&C Printing Co., Ltd.
Printed in China

ISBN 978-3-7913-8329-3

Front cover: Head and neck showing muscles of the cheek, oral cavity and pharynx, 1893, Ruge Collection

Museum Vrolik is the anatomical museum of the University of Amsterdam. It has been located in the Academic Medical Centre of Amsterdam since 1984. The museum is named after father and son Vrolik, who were both professors of anatomy in nineteenth-century Amsterdam. Their 'Museum Vrolikianum' was a private collection of human and animal anatomy that relocated to the University of Amsterdam after their deaths. Later anatomists expanded the collection, but it has remained known as 'Museum Vrolik'.

Museum **Vrolik**

/museumvrolik

Museum Vrolik
Academisch Medisch Centrum
Meibergdreef 15, J0-130
1105 AZ Amsterdam